R-ticulate

R-ticulate

A Beginner's Guide to Data Analysis for Natural Scientists

Martin Bader
Department of Forestry and Wood Technology
Linnaeus University
Växjö, Sweden

Sebastian Leuzinger
School of Science
Auckland University of Technology
Auckland, New Zealand

Published by John Wiley & Sons, Inc., Hoboken, New Jersey.
Published simultaneously in Canada.

For general information on our other products and services or for technical support, please contact our Customer Care Department within the United States at (800) 762-2974, outside the United States at (317) 572-3993 or fax (317) 572-4002.

Wiley also publishes its books in a variety of electronic formats. Some content that appears in print may not be available in electronic formats. For more information about Wiley products, visit our web site at www.wiley.com.

Library of Congress Cataloging-in-Publication Data

Names: Bader, Martin (Professor), author. | Leuzinger, Sebastian, author.
Title: R-ticulate : a beginner's guide to data analysis for natural scientists / Martin Bader and Sebastian Leuzinger.
Description: Hoboken, New Jersey : Wiley, [2024] | Includes index.
Identifiers: LCCN 2024001120 (print) | LCCN 2024001121 (ebook) | ISBN 9781119717997 (cloth) | ISBN 9781119718000 (adobe pdf) | ISBN 9781119718024 (epub)
Subjects: LCSH: Science–Data processing. | Science–Statistical methods. | R (Computer program language)
Classification: LCC Q183.9 .B33 2024 (print) | LCC Q183.9 (ebook) | DDC 502.85/5133–dc23/eng20240215
LC record available at https://lccn.loc.gov/2024001120
LC ebook record available at https://lccn.loc.gov/2024001121

Cover Design: Wiley
Cover Image: © Martin Bader

Set in 9.5/12.5pt STIXTwoText by Straive, Chennai, India

Contents

Foreword

It has been over 30 years since Robert Gentleman and I began work on what would eventually become "R". At the time, we had rather modest goals; we thought that we might be able to use the software to support our own teaching and perhaps some research.

The development of the concept of free software and the GNU system in particular was attractive to us from both philosophical and practical viewpoints. Our decision to release R as free software using the GNU license created by Richard Stallman was fateful. To our surprise, the software seemed of interest to a large number of talented developers who quickly became collaborators. The work of this group was instrumental in creating a platform that appealed to both users and developers.

Despite its now venerable age, R remains a useful tool. Thousands of add-ons have been created that have moved R into areas of application that we could not have envisaged when we started work on it. This book is an example of this kind of use. It shows that R is suitable as a companion software for those students who major in subjects other than statistics and mathematics.

If you put R to use, please do so in the spirit that it was created. Collaboration is a very powerful tool, and freely sharing the results of your work creates benefits for all, including you.

January 2024

Dr Ross Ihaka
Co-founder of R, Auckland

You can't have too many stats books – at least that seems to be the message from my office book shelves. For a mathematical subject, statistics is surprisingly philosophical and multi-faceted – rich enough to justify many views and presentations. And then there's the issue of how to implement the chosen analysis. Here, Bader and Leuzinger provide a non-technical, reader-friendly introduction to linear model analysis using R. An attractive feature of the book is the extensive use of explanatory boxes and the demonstration of key ideas (e.g. pseudoreplication) through simple simulations. The 12 chapters provide an accessible introduction to basic statistical principles, all while providing the keener student with a glimpse of the more advanced methods. The authors' approachable and dynamic style makes this textbook a gem for students and teachers who are seeking quick access to modern statistical methods.

February 2024

Prof. Andy Hector
University of Oxford, UK

Preface

The way statistics is taught in undergraduate university courses for natural scientists has changed enormously over the past two decades. On the one hand, the use of a software accompanying both theory and practical sessions has become the standard, facilitating access to an unprecedented and ever-increasing wealth of statistical methods. On the other hand, the ease with which some of these more complex methods can be applied has meant that the theoretical understanding of what is going on under the hood has suffered. In teaching undergraduate statistics over many years, we have found that there is a danger of skipping some of the basics and heading straight to the strategy of 'googling – copying code – trying to apply it to my data', or, more recently, letting AI do the heavy lifting. This may result in an uncritical and sometimes incorrect use of statistical techniques. We seriously doubt that one can ever understand variance modelling or even a simple non-parametric test without a fundamental comprehension of the basic concepts of distributions, test statistics, but also more broadly applicable ideas such as the principle of parsimony. In our teaching experience, we found that there is a gap between introductory texts that have encyclopaedic character and lack conciseness and the myriad of more specialised texts that are clearly out of reach for an undergraduate student.

Our aim was to produce a surmountable, paced, 12-chapter text which suits the typical 12 week (1 semester) layout of many undergraduate courses. Just like in our classroom teaching, we use easy-to-understand language to explain key concepts and strive to strike a balance between the practical application of statistical models and a conceptual understanding of the theoretical foundations. The content of the chapters can be used selectively, and we deliberately go deeper where we deem detail is useful. As such, the text may in part also suit postgraduate students of the natural sciences. Our philosophy generally suits the curious student's attitude of 'wanting to know why', but we make sure we summarise the key concepts for those who are after a simple 'recipe' to get there. Margin texts and dedicated boxes will help with this. We demonstrate the use of base R as well as that of more recent developments (tidyverse) as we firmly believe that both have their place, and taking the best of both worlds makes most sense. Exercises and datasets are available online to save space and keep the volume of the text 'light', particularly for those who prefer to work with a hard copy. We hope this text will help you to *R-ticulate* yourself well in a data-driven world where statistical analysis and modelling skills are ever more important and in high demand with employers, far beyond the natural sciences.

January 2024

Martin Bader
Växjö, Sweden

Sebastian Leuzinger
Auckland, New Zealand

About the Companion Website

This book is complemented by a companion website.

www.wiley.com/go/Bader

The companion website provides all datasets used in the book, as well as exercises with solutions for each chapter.

1

Hypotheses, Variables, Data

In a world where information circulates at unprecedented speed and almost exclusively via the internet, often on indiscriminate platforms, it has become increasingly difficult to distinguish between *fact* and *fake*, between *true* and *false*, and between *evidence* and *opinion*. One admittedly non-spectacular, yet indispensable way to confidently plough our way through the jungle of those dichotomies is learning and understanding the basics of statistical thinking. Thinking statistically needs training, as it is not intuitive. Humans are particularly bad at 'collecting' data. For example, the perception of whether we had a 'warm' or 'cold' winter will depend on a wealth of subjective factors such as how much time someone has spent outdoors, and likely shows little correlation with the actual average temperature of that particular winter. Similarly, people perceive risks in life in an utterly 'non-statistical' way. While, statistically, the largest risks for early death in most western countries are sugar intake and lack of exercise, we often perceive the risk of an airplane crashing or a great white shark attack as much more threatening to our lives. In fact, the mentioned risks are four to five orders of magnitude (about 100,000 times) apart!

Human perception is a notoriously bad statistician

In this introductory chapter, we will set the foundations for 'statistical thinking', and the most basic statistical skills, which are the pillars that scientific thinking in a broader sense rests on. From our own experience, confusion at a later stage of someone's scientific career is often caused by a lack of knowledge of the fundamentals of statistical principles. For example, we often get approached by students asking us for statistical help, but then the initial conversation shows that the student is not clear about how many variables they are looking at, what their nature is, which of them might be a response or a predictor variable, what the unit of replication is, and so forth. Confusion can also originate following data collection in the absence of a sound scientific hypothesis, or from the absence of a sound study design. Pretty much everything is bound to go wrong from there if these foundations are not laid in time. The purpose of this chapter is therefore to introduce a basic statistical vocabulary, including the formation of a good scientific hypothesis, and how it relates to your data. This also includes training ourselves to dissect and categorise the datasets we encounter. We need to clarify the number of variables we are looking at, as well as their nature and purpose in the dataset.

Alongside all this, we will slowly ease you into the use of the statistical software 'R'. Despite this adding an element of complexity, it is useful to learn the theory at the same time as the use of a software. We recommend that you replicate our examples on your computer to get a hands-on experience.

R-ticulate: A Beginner's Guide to Data Analysis for Natural Scientists, First Edition.
Martin Bader and Sebastian Leuzinger.

Companion website: www.wiley.com/go/Bader

1.1 Occam's Razor

The principle of parsimony (or Occam's razor) is an extremely useful one that should accompany us not only in our statistical thinking, but also generally in all our scientific work. The idea goes back to William of Ockham, a medieval philosopher, who articulated the superiority of a simple as opposed to a more complex explanation for a phenomenon, given equal explanatory power. An illustrative example is the heliocentric model (the earth rotating around the sun) versus the geocentric model (which has the Earth in the centre of the solar system). Both are (or were) used to explain the trajectories of objects in the sky. It turns out that the heliocentric model is far simpler with equal or superior explanatory power, and is thus preferable. The search for the simplest possible explanation has been at the origin of many scientific discoveries, and it is useful to let this principle guide us throughout all stages of planning and conducting our data collection and analysis. The iconic phrase 'It is in vain to do with more what can be done with fewer' is applicable to the various stages of scientific research process:

'It is in vain to do with more what can be done with fewer'

(1) Planning stage. What is our hypothesis? What data do we need to answer the question? The clearer our (scientific) question is formulated, the easier it is to decide what kind of data we need to collect. 'De-cluttering' the relationship between the question we want to answer and the data we collect can save us an enormous amount of time and frustration.
(2) Data cleansing and organising stage. Remove any unnecessary element in our data, use minimal (but unique) nomenclature, both for variable names and values. A well organised and simplified dataset forces us to gain a much better understanding of it.
(3) Analysis stage. Use the minimum number of explanatory variables (and their interactions) that best explains the patterns in your response variables. Every additional explanatory variable will explain some variation in the response by pure chance. We will get to know statistical tools that help us decide on the most *parsimonious* model, i.e. the smallest set of explanatory variables that explain the most variation in the response variable(s).
(4) Presentation stage. Once we are ready to convey our (statistical) findings, we again use 'Occam's razor' to reduce and 'distill' our texts, figures, and tables. This means the removal of any frills that do not serve to make our results more comprehensible. Often, information (for instance, graphs) can be condensed, white space can be minimised or used for insets. In short, we maximise the ratio of information conveyed over the space used (see Chapter 6).

Maximise $\left\{\frac{\text{Information}}{\text{Space, Time}}\right\}$ and $\left\{\frac{\text{Explanatory power}}{\text{\# of explanatory variables}}\right\}$

We will revisit most of the aforementioned strategies in chapters to follow. The guiding principle can be extended to many more areas of science not covered in this book, such as scientific writing and oral presentations.

1.2 Scientific Hypotheses

Sometimes we are confronted with complete datasets that we obtain from somewhere, which we then have to analyse. Other times, we pose the scientific question ourselves, and we then set out to collect the data. In both cases, we need to be absolutely clear what question we can or want to answer, as this question is linked with our *scientific hypothesis*. A scientific hypothesis is a well-founded assumption that must be testable. For example, 'The majority of consumers under 20 years of age prefer to pay cashless', serves as a scientific hypothesis, while 'Young people prefer to pay by card' lacks the level of precision required for a scientific hypothesis that we can test, but it may serve as an idea that could lead to a scientific hypothesis. A mismatch between an early (often vague) idea of what we want to research, the resulting hypothesis, and the data we set out to collect, can lead to much confusion.

Be absolutely clear what question you want to answer, and how it could be turned into a hypothesis!

Say, we are interested in biodiversity changes in rock pools on a rocky shore along a tidal gradient. This initial idea could lead to various hypotheses, which would require different data collections. If our hypothesis is 'The closer the rock pools to the high tide mark, the fewer species per pool', we would want to count the number of species per rock pool. However, if our hypothesis is 'species A, B, and C become more abundant moving from the low to the high tide mark, while species D, E, and F become less abundant', then we would count the number of individuals per species and rock pool. The hypothesis 'Biodiversity decreases along a tidal gradient' may not be precise enough to decide exactly what kind of data need to be collected. It is always advisable to have a clear idea of the units we are using (also referred to as the 'metric', or the 'reference metric' to refer to the denominator of the metric only). For example, if we count the number of species *per pool*, we will get a fundamentally different outcome compared to if we had counted the number of species *per cubic metre of water*, or *per square metre of rock pool surface*. In some instances, the choice of the reference metric can completely change the meaning of the data: while the 'cases of Covid-19 per 1000 people' in a country can increase from one week to another, the 'cases of Covid-19 per 1000 administered Covid tests' can decrease at the same time. These examples illustrate how quickly confusion can arise, which can lead to misinformation, but also how easily a statistically uninformed audience can deliberately be misled. Having a crystal clear idea of what the research question is, what hypothesis we test, and what data we need to collect is therefore pivotal not only for our own work, but also to scrutinise and criticise data or findings that are presented to us.

What units are my variables in? What reference metrics am I using?

The aforementioned considerations are important, but so far they are purely theoretical. To really understand and 'experience' these issues, we need to 'leave the classroom' and get our hands dirty!

1.3 The Choice of a Software

This topic can be delicate, as people's beliefs, experiences, preferences, and even commercial interests play a role, such that the choice of a statistical software is ultimately a very subjective decision. We will spare you with all the arguments that could be used to advertise R. It is important to note that the R project is entirely non-commercial, so the software is absolutely free. R is easy to download on all major operating systems. We additionally recommend downloading the user interface 'RStudio', which is also free for non-commercial individual use. With easily accessible online help, there is not much more to instruct here than telling you to install R (by searching for 'R software' and then clicking your way through), and installing 'RStudio' in the same way. You should then always open 'RStudio' rather than R, as 'RStudio' will run R in the background.

1.3.1 First Steps in R

Open 'RStudio'. Go to file -> New file -> R Script (basically just a text file with your code). Figure 1.1 illustrates what RStudio will look like. Focus on the left two windows, the top one

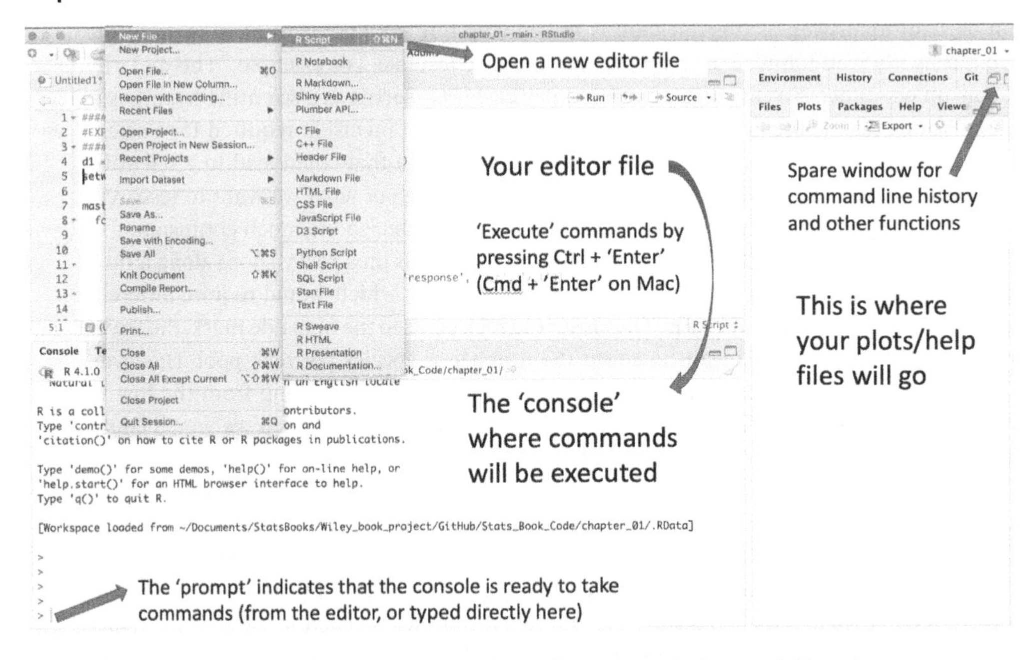

Figure 1.1 RStudio, showing the editor file (top left), the console (bottom left), and two extra windows to the right for plots, help, etc. (the top one has been collapsed here).

Send code from the editor to the console by pressing CTRL + Enter (PC) or Command + Enter (Mac)

is the editor (where your code will go), and the bottom one is the so-called 'console', where code is evaluated. That means that at the top, you can write whatever you like, R will not care until you *evaluate* or *run* it, by sending the code to the console. To illustrate this, type '1 + 1' into the top window (the editor), and press enter. You will see that nothing happens, except that the cursor jumps onto the next line, just like in a text document. Do the same thing in the console – and you will see the result (2) appear, and the prompt ('>') pops back, awaiting the next task. We will see what the '[1]' that you see in front of the result means later. As a beginner, you will often make typos, forget brackets, etc., causing R to get stuck or hung up, which you notice by the missing prompt sign in the console. Instead, you usually see a 'plus' sign in the console because R awaits further input (e.g. missing brackets). Often, it is difficult to figure out where things have gone off the rails. If this happens, simply hit the ESCAPE key, which forces R to terminate the incomplete code and the prompt should pop back up.

Mathematical operators:
* times
/ divided by
() brackets
^ to the power of
sqrt() square root
exp() exponent
log() logarithm

The best way to become familiar with the use of R is to simply play around. Much of what follows sounds more complicated than you will find once you get your hands 'dirty'. Start with playing in the console, for instance, by typing mathematical operations. You can do the same thing in the editor of course, and then 'send' your commands to the console to be evaluated. For example, type '10-7' into the editor, and then simultaneously press 'Ctrl' and 'enter' ('Cmd' and 'enter' on Mac). The cursor will stay up in the editor window, but you will see the result in the console, and you are ready for the next command to be evaluated. See margin for a number of mathematical operators that R can interpret and you can play with, e.g.:

```
> ## R as a calculator
> 1 + 2 * 3
[1] 7
```

```
> 2 * 5^2 - 10 * 5
[1] 0
> 4 * sin(pi / 2)
[1] 4
> 0 / 0   #not defined (Not a Number)
[1] NaN
```

R is also a powerful calculator!

The hashtag (#) allows you to leave comments in your code – anything after the hashtag is ignored by R!

So when should you type into the editor, and when directly into the console? Whatever you type into the console is not saved, but code that is typed into the editor file will be saved. So use the console to try things out, and maintain a clean editor file with code that works, and that you want to save. It is advisable to make sure the entire code runs if you send it down to the console, i.e. you do not receive any error messages. Here is a list of tips and tricks that make your start with RStudio easier:

- If you want to evaluate several lines of code that you have written in the editor, highlight the part of code you would like to evaluate, then press 'Ctrl' and 'enter' (simultaneously, henceforth referred to as *evaluating* or *running* code); partial code evaluation also works when you highlight the desired bit of code with the cursor, i.e. you can execute a command that is wrapped inside one or more commands. Very helpful when studying complex code examples!
- If you press 'Ctrl' and 'enter' with no code highlighted, the whole line that the cursor sits in will be evaluated (regardless of the exact position of the cursor).
- If you would like to leave a comment that R should ignore, place a hashtag in front of it, e.g. 7*52 # days per year. R will then ignore any characters after the hashtag in that particular line, so it will calculate 7×52, but ignore the comment 'days per year'.
- If you are typing directly into the console, use the 'up' arrow to retrieve previously typed lines. This saves you a lot of typing!
- When you close RStudio, you will be asked which files you want to save. It is critical to save the editor file (essentially a series of instructions on what R is to do with your raw data). The 'workspace' contains the actual objects that are created, resulting in a relatively large file size. Saving the 'workspace' is much less important, as you will be able to recreate it easily with your saved code (as an editor file a.k.a. R script, usually with the suffix .R). It is only worthwhile saving the workspace if you have run some models or algorithms that took substantial computing time. In these cases, opening a saved workspace allows you to pick up things where you have left them.

Use the 'up' arrow to retrieve code in the console

1.4 Variables

The ability to develop a fundamental understanding of what variables we encounter in a dataset is at the onset of any data analysis. For example, if we cannot answer with confidence how many variables we are looking at, what distribution their values follow, whether they are likely predictor or response variables etc., then we are in for a lot of trouble. Some of the following points might appear trivial, but from our experience, it is important to go through these notions in detail in order to build on solid grounds.

1.4.1 Variable Names and Values

Whenever you come across data for the first time, try to see them in variable names and their values. This is a bit like sorting out a pile of matches – get all the heads (variable names) lined up with their wooden ends pointing downwards (their values). We could, for example, have three variables in a dataset: age, height, and weight (those are the variable names). If we have four observations (values), those might be [41, 38, 73, 29] for the variable 'age', [178, 163, 159, 181] for the variable 'height', and [82, 78, 59, 92] for the variable 'weight'. It seems simple and easy to see those three matchsticks line up in front of our imaginary eye.

Use '=' or '<-' to assign values to objects. They are equivalent. The shortcut 'alt' + 'dash' (minus key also works) should be used to efficiently type the assignment 'arrow'

To make this a little less theoretical, it is best to set up these variables in R:

```
> ## Assigning variables
> age <- c(41, 38, 73, 29)
> height = c(178, 163, 159, 181)
> weight <- c(82, 78, 59, 92)
```

The `c` is an R function (see Box 1.1), just like `sqrt` used earlier. It stands for 'concatenate', so we are asking R to arrange the values in the parenthesis in the form of a variable. Note that the '=' sign is equivalent to the frequently used 'assignment arrow' ('<-'). We can now collate those variables into a dataset. R calls this a 'data frame':

Use '`data.frame`' to create a dataset from variables

```
> d1 <- data.frame(age, height, weight)
> d1 # show the dataset
  age height weight
1  41    178     82
2  38    163     78
3  73    159     59
4  29    181     92
```

When assigning objects as earlier, note that the names (age, height, weight, d1) can be chosen freely. We use 'd1', meaning 'dataset 1', as often we make changes and end up with a modified dataset, which we can then call 'd2'. Also note that unless you subsequently type the object name (e.g. 'd1') into the console, you will not see the dataset, as it is simply assigned to the object name you have chosen. Make sure you do not overwrite existing function names such as `data` or `df`, as tempting as it may be. The command `data` (used without any argument) shows the built-in datasets and those available in loaded R packages. It is also used to load data from packages (that do not come with the base installation) into the current R session. The command `df` is the density function of the F-distribution. R will not get corrupted when you assign an object to an existing function name, but such stunts may entail some funny behaviour down the track.

Before assigning an object to a name, make sure the name is available in R and not already used as a function name or in-built object

Box 1.1 Functions in R

The syntax of R is simple in the sense that whatever you do, you are mostly calling functions, i.e. you type the name of the function (e.g. `round`), open a parenthesis, type the arguments that the function requires separated by commas, and close the parenthesis. For example, the function `round` requires 2 arguments: (i) a number or an entire variable whose values are to be rounded, and (ii) the number of decimals desired. In the round function, the second argument has a default value (zero), i.e. if you do not specify it, R will leave zero decimals to the number(s) you want to round. RStudio will display what arguments are required at the bottom of the console or right when you start typing a function (this is an RStudio feature and is called 'argument preview'; in some cases, you need to hit the 'tab' key to see the preview). Note that as long as you are aware of the order in which the arguments are provided by default, you do not need to name them. If the argument name is unique, you do not even need to spell it out. The following example shows this:

```
> anything  <- 1.1928
> round(digits = 2, x = anything) # default order reversed by correct
argument naming
[1] 1.19
> round(di = 2, x = anything) # unique abbreviation of 'digits'
[1] 1.19
> round(anything, 2) # no argument names, but correct order
[1] 1.19
> round(2, anything) # no names, wrong order - not what we want!
[1] 2
```

If you are not sure what or how many arguments you have to provide to a function, simply type a question mark followed by the name of the function you want to learn more about, execute the code, and a help file will open (normally to the right of the console). For example, to access help for the function `mean` type '?mean' straight into the console. Under 'Usage' you will see all arguments with their default values (if they exist) indicated after the '=' sign. You must specify arguments with no default values or R will complain. This makes sense: for example, if you use the function `round`, it is reasonable for R to assume that you want to round to a whole number. Therefore, the argument `digits` is set to zero by default. However, there is obviously no default value for the argument `x`, the actual number(s) you want to round. In the help file, under 'Arguments', you will find a detailed explanation of each argument, and under 'Value' you are explained what the output of the function is (for example, the rounded numbers in the case of `round`).

Of course you can nest functions, e.g. you can round the logarithms of some numbers:

```
> round(log(c(1, 2, 3, 4, 5)), digits = 1)
[1] 0.0 0.7 1.1 1.4 1.6
```

The following are some more examples of functions that you might find useful, both to practice but also for data preparation. You will probably figure out what these functions do and what arguments they require.

```
> max(c(0, 4)) # find the maximum
[1] 4
> rep(1:3, times = 2) # repeat 1,2,3 two times
[1] 1 2 3 1 2 3
> seq(from = 0, to = 1, by = 0.1) # sequence from 0 to 1 spaced by 0.1
 [1] 0.0 0.1 0.2 0.3 0.4 0.5 0.6 0.7 0.8 0.9 1.0
```

This was simple, but what if things get a bit more complex? Consider this example: a student records the time anemone fish spend guarding, hiding, or foraging. The study compares behaviour in captivity vs. in the wild, where the observations are carried out at 1 and 3 m depth. This is a much messier pile of matches! What are the variable names? What are their values? If 'Time' represents a continuous one-second observation interval, a possible answer could be:

Time	Behaviour	Depth	Site	Species
1	Foraging	1	Aquarium	A
2	Foraging	1	Aquarium	A
3	Guarding	1	Aquarium	A
4	Hiding	1	Aquarium	A
...	...	...	...	...

In this case, we have five variables (time, behaviour, depth, site, and species). We could also organise the dataset to represent the variable 'time' as the cumulative time in seconds that a fish spends performing a certain behaviour. In this case, the values of the variable 'time' are no longer a continuous representation of time, but accumulated time spans (e.g. 2 seconds spent foraging, then 1 second spent guarding, etc.):

Time	Behaviour	Depth	Site	Species
2	Foraging	1	Aquarium	A
1	Guarding	1	Aquarium	A
1	Hiding	1	Aquarium	A
...	...	...	...	...

Yet another way of organising the same dataset would be to assign TRUE (T) and FALSE (F)

Time	Guarding	Hiding	Foraging	Depth	Site	Species
1	F	F	T	1	Aquarium	A
2	F	F	T	1	Aquarium	A
3	T	F	F	1	Aquarium	A
4	F	T	F	1	Aquarium	A
...	...	...	...	...	...	...

We will come back to data formats in the section on the wide vs. long format (see Section 1.5.1), as these have implications for the subsequent analyses. For now, it is important that we develop a skill to put order into our heap of matchsticks – i.e. given an already collected dataset, or in view of a planned data collection, we must be very clear about what variables we collect (their names) and what potential observations there could be (their values). This will be an important prerequisite for anything that follows.

Creating this dataset in R is similar to what we have done earlier. Here, we also need to create variables whose values are not numbers, but characters, so, for example:

```
> site <- c('Aquarium', 'Aquarium', 'Ocean', 'Ocean')
> site
[1] "Aquarium" "Aquarium" "Ocean"    "Ocean"
```

For larger datasets, you are more likely to import them as a spread sheet or text file (see also Box 1.2: 'Reading data into R').

Box 1.2 Reading data into R

You will likely enter data either into a text editor or into a spreadsheet software. If you use a text editor for data entry, separate individual data points by tabs and then use `read.table`. If you enter data via a spreadsheet, make sure not to use any special formatting, and all rows and columns have the same lengths. Do not use multiple worksheets, they will be lost upon importing into R. Save your file as .txt (tab delimited) or .csv (comma delimited). Then use `read.table` or `read.csv`. You can use header names in the first row, as the argument 'header' is set to TRUE by default in `read.csv` (but not in `read.table`, see the following code). The main difference between `read.table` and `read.csv` is that `read.table` also reads row names, often causing the header row to be one element shorter than the rest. Using `read.csv` may also be safer against the common trap of using spaces in the header names (e.g. 'plant weight'), which R would interpret as two separate entries. If the rows or columns are not of equal length, R cannot produce a data frame – a common mistake if you do not check your spreadsheet thoroughly first. The same happens if you have blanks for missing data – replace those with 'NA' before you import the data, as R will recognise these as missing entries. If your file sits in the same folder as your workspace is pointing to, you do not have to specify a path:

Use `read.table` or `read.csv` to read data into R or `read_xl` (package *readxl*) for Excel files, which allows the user to specify the sheet to be read

```
> d1 <- read.table("dataset1.txt", header = T) #(tab delimited, with
  headers)
> d1 <- read.csv("dataset1.csv") #(comma delimited, with headers)
```

The aforementioned code will only work if R is looking in the right folder, i.e. the one in which the desired file is located. This folder is called the 'working directory'. You can think of the working directory as your 'target folder', where R assumes to find files you intend to read in and where R exports saved objects such as graphs, data or model outputs to (similar to the 'Save as...' option in MS Office software allowing you to save files in a specific location). It is best practice to set the work directory right at the start of a new R session. This is most conveniently done via the RStudio menu bar at the very top (Session -> Set Work Directory -> Choose Directory). With this approach the command for setting the working directory (`setwd`) shows up in the console from where you can copy and paste it at the start of your R script, so you automatically point R to the correct folder next time you run your R script:

```
> setwd("C:/...path.../myProject") # adjust the path to your system!
```

The two commands `file.choose` and `list.files` can help a lot if you do not know what the path to your file is, or whether you have your raw data in the correct folder. Use `file.choose` (with no arguments, so just type `file.choose()`) for a popup window which allows you to select a file, whose file path then gets printed into the console. The `list.files` function allows you to view all files in your workspace. For example, if `read.table` does not work, type `list.files()` into the console to see if the file you want to read actually sits in your workspace. It is quite common not to succeed with reading data into R at the first try, because of a mismatch between file location and working directory. You will soon figure out what works best for you, given your operating system and spreadsheet software.

(Continued)

Box 1.2 (Continued)

If you are reading more complex files, for example, with some header that you want to exclude, both `read.table` and `read.csv` offer a wealth of arguments to customise reading your files into R, for example, you can use the argument '`skip = x`' to disregard the first x lines in a file (possibly because there are meta data that you would want to exclude). Other useful arguments are '`na.strings =...`', which allows you to specify how missing values are dealt with. Generally, R refers to missing data as NA (not available), but you might have different indicators for missing data in your raw file. For example, if both periods and blanks indicate missing data, we can use '`na.strings = c(" ", ".")`' to turn those observations into NAs during data import. Setting the argument `strip.white = TRUE` automatically removes leading or trailing white space, which may create spurious factor levels (e.g. 'alive' is not the same as 'alive ' – note the trailing white space in the second occurrence, leading to two factor levels for that variable).

1.4.2 Types of Variables

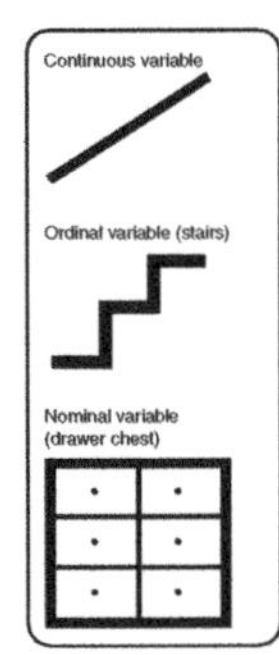

Understanding what types of variables we have in our dataset will help us later on with our analyses, but it also provides us with the necessary statistical vocabulary needed to read and communicate statistical results. The fundamental division into continuous and categorical (sometimes called discrete) variables is most important, as very different statistical methods are used for those types of variables (Table 1.1). A continuous variable can take on any numeric value, while a categorical or discrete variable has two or more categories (levels).

Table 1.1 A brief overview of variable types. The term 'factor' is sometimes used synonymously with 'discrete variable', but occasionally also to refer to a continuous variable, for instance, when 'speed' in a dataset is referred to as a 'risk factor'. Some authors use the term *categorical* synonymously with *nominal*, and define discrete variables as numeric variables only.

Variable	Explanation	Examples
Continuous	A variable that can take on any value between its minimum and maximum value	Height, distance, mass, speed, temperature
Categorical/discrete		
Nominal	A variable that has two or more categories that cannot be ordered	Habitat type, colour, species
Binomial	A categorical variable with only two categories that cannot be ordered	Presence/absence, dead/alive
Ordinal	A variable where the categories can be intrinsically ordered	Fertiliser application (low, medium, high), age classes, rating scales
Counts	Whole number values (integers). The number of times an event or outcome occurs	Species abundance, particle counts, material failure events, disease frequency

Body height, for example, is a continuous variable, while marital status is categorical (you can only be one of 'single', 'married', 'separated', 'divorced', or 'widowed', i.e. there is no possible value in between 'married' and 'widowed'). The same applies to continuous variables that are binned to form discrete categories (e.g. income or age bracket). We further distinguish variables that can only take on two possible values (such as dead/alive, or smoker/non-smoker), we call those *binomial* (or *binary*).

Ordinal variables are a type of discrete variables where there is a clear intrinsic ordering of levels. For example, suppose you are investigating invasive species whose dispersal potential is described as low, medium and high. You may not be able to put a number on these levels, but as their names suggest, they can be ordered from low to high. Another example are rating scales such as the classical 5-point Likert scales (e.g. *Strongly Disagree – Disagree – Undecided – Agree – Strongly Agree*) (see Table 1.1).

In R, every variable, and more broadly every object in general, will be of a certain class. For example, a continuous variable will be of class 'numeric' a nominal variable of class 'character', and a data frame will be of class 'data.frame'. Even a model output will be assigned a certain class, for an overview of object classes, see Box 1.4. 'R Objects'.

Continuous vs. Categorical variables

1.4.3 Predictor and Response Variables

Once we have organised our dataset such that we are clear about *how many* variables we have, *what their names are*, *what (potential) values they could take on*, and *of what nature* they are, we can identify which variables are used to predict patterns in other variables ('predictor variables'), and which variables are *being predicted* ('response variables'). There is rarely a definite classification, often a variable in a dataset can take on the role of a predictor *or* a response variable. To illustrate this, the binomial variable 'smoker' (with values 'yes'/'no') and the variable 'yearly income' could both be either response or predictor: smoking habit could influence yearly income (if you are a smoker, you likely earn less), or vice versa (if you earn less, you are more likely a smoker). The decision which variables are used as predictor or response variables is often not trivial and rests on good knowledge of the subject matter. In some cases however, particularly in planned experiments, the classification into predictor and response variables is much easier. For example, in a medical trial, the drug dose administered (e.g. 2, 5 or 10 mg) is clearly a predictor variable, while the patients' response (e.g. the resulting heart rate) will be a response variable. Some textbooks refer to predictor variables as 'independent variables', as (at least in an experimental context) they are independent, i.e. in the aforementioned example, *you* decide on the dose rather than it being an experimental outcome. Conversely, response variables are 'dependent' on the predictor(s), and thus referred to as 'dependent variables'. Yet another common name for predictors is *explanatory variables*. Here, we will stick to the intuitive predictor/response nomenclature.

Predictor variable = explanatory variable = Independent variable; Response variable = Dependent variable

We will always consider one response variable at a time. If several response variables are considered simultaneously, we need *multivariate* methods that are not discussed here. Often, we perform similar tests on a dataset, subsequently using different response variables. Conversely, we can use multiple predictor variables at the same time, each of them explaining some variation in the response variable.

1.5 Data Processing and Data Formats

Everything we have discussed so far is essential. Moving on without the confidence that we can name our variables, characterise them, identify predictor vs. response variables would not be wise. Before we dig further into our statistics toolbox however, we have to acquire some skills that we often refer to as 'data crunching'. We might need to merge certain variables, transform them, add new ones, remove outliers, etc. We are guiding you through a condensate of the tools we have personally found most useful over the years, in both research and teaching. R has expanded so rapidly that for every problem you encounter, particularly while processing data, there is a wealth of different solutions, sometimes making it difficult to decide which one to aim for. Sometimes the sleekest way to do the job is not the most intuitive, and you may want to go the longer, but more understandable way. Downloading and using a dedicated R package to solve a certain problem may be useful if you do that same job often. For a one-off application however, you may choose to use base-R functions, or even solve issues in a spreadsheet software, before importing the data into R.

1.5.1 The Long vs. the Wide Format

The matchstick analogy used earlier is again useful when we think about organising our datasets. If our data already are in this format, i.e. all variable names are in the first row, then we are fine. Often however, the most practical way to collect data or the format data are issued from an automatic device such as a data logger is different. Imagine you are collecting blood count data from five different mice every week over three weeks. Likely, you will have one column per week, and five rows for the five mice (Figure 1.2). Here, the variables are not 'matchsticks', because your response variable (blood count) is spread over three different columns. If you would like to organise this dataset in long format, then you will need the headers 'mouse', 'week', and 'score' as variable names, as shown in Figure 1.2. The values of 'mouse' and 'week' will obviously have to be repeated three times, increasing the total volume of the dataset. You will agree that the long format is more generic – there is only one way to organise your data like this, while the wide format leaves many options, e.g. you could have the five mice in a column each, and the three weeks as rows 1–3. Toggling between the two formats can be done outside R in your spreadsheet software, or you can use the function `reshape` in R, which we will not discuss further here.

Use the long format, the heads correspond to variable names

It is now easier to appreciate why carefully identifying/characterising our variables, and then organising them in the long format are two inseparable and fundamental processes, where there are no shortcuts. There are a few cases where we may want our data in a different format, suitable for specific tests (e.g. a contingency table to perform a Chi-squared test, see Section 7.3 'The Chi-squared Test'). It is always easiest however to start from the long format and produce other formats from there.

1.5.2 Choice of Variable, Dataset, and File Names

The earlier mentioned 'Principle of parsimony' or 'Occam's razor' (see Section 1.1) should be considered when choosing any variable, dataset, and even file names. For variable names,

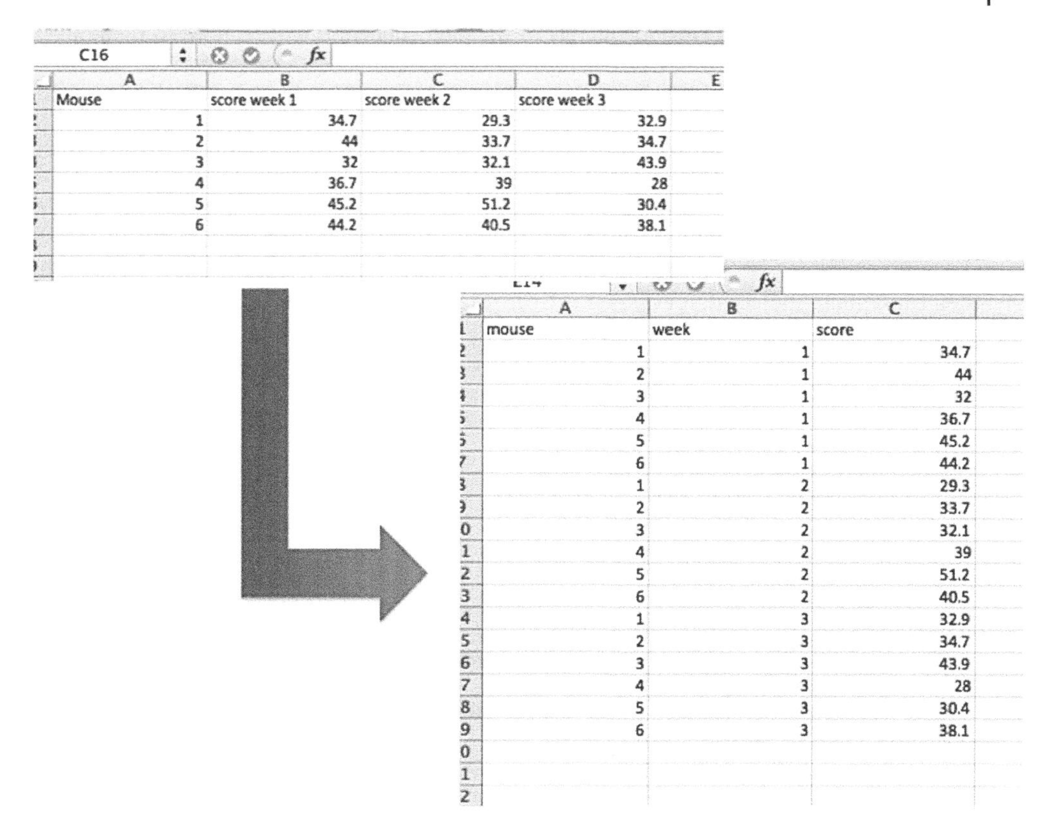

Mouse	score week 1	score week 2	score week 3
1	34.7	29.3	32.9
2	44	33.7	34.7
3	32	32.1	43.9
4	36.7	39	28
5	45.2	51.2	30.4
6	44.2	40.5	38.1

mouse	week	score
1	1	34.7
2	1	44
3	1	32
4	1	36.7
5	1	45.2
6	1	44.2
1	2	29.3
2	2	33.7
3	2	32.1
4	2	39
5	2	51.2
6	2	40.5
1	3	32.9
2	3	34.7
3	3	43.9
4	3	28
5	3	30.4
6	3	38.1

Figure 1.2 The wide vs. the long format.

choose the simplest, unique name that contains as little 'clutter' as possible. 'pd' is better than 'PD', which is better than 'percent_damage', which is better than 'percent damage', which again is better than '% damage'. Firstly, keep variable names short as you will have to type them often. Avoid hard to type underscores, periods, and capitals (R is case sensitive, and you will forget if you capitalised a variable name, so stick to lower case where possible). Spaces are a no-go, as this is an easy way to confuse both R and yourself. Special characters are probably the worst, as they often mean something in the R syntax. For variable values, use numeric values where possible, e.g. for the variable 'age class', do not use 0–30, 31–40, etc., but rather code them as age class 0, 1, 2, ... This makes it much easier later to work with the dataset, for example, if you would like to filter out age classes 0 and 1. If a numeric variable contains one single 'messy' entry, for example, '>200' signifying 'greater than 200', R will interpret the entire variable as text (class 'character'), which will most certainly lead to confusion. Such 'data cleansing' is important to achieve a clean dataset, and should be done at the earliest possible after data collection.

Keep variable names as simple as possible, avoid spaces, capitals, and special characters

The nomenclature of datasets and files is perhaps less critical, but even here, it pays to be parsimonious and consistent. Always keep your original raw dataset in a safe place so you can go back to it. Including dates and unique identifiers are equally important. A lot of information that is obvious on the day of the data collection will be hard to remember later.

1.5.3 Adding, Removing, and Subsetting Variables and Data Frames

If you only want to use a subset of your data, you have to indicate your selection in (square) brackets after the variable or data frame you want to subset. You can either specify the values you want to select (the 4th to the 6th: variable [4 : 6]), or use logical expressions, which then cause those indices that evaluate to 'true' to be displayed. In practice, you will mostly subset data frames and variables, but you can apply similar methods to lists and three-dimensional arrays, which we will not discuss here. We will first look at how to subset a single variable.

```
> x <-  c(1, 2, 3)
> y <-  c(2, 2, 2)
> y[1] # element 1 from y
[1] 2
> y[1:2] # elements 1-2 from y
[1] 2 2
> y[2:length(y)] # second entry to last (length(y) evaluates to 3)
[1] 2 2
> x[1:2][1] # double indexing (first element of a subset variable)
[1] 1
> x[x > 1] # all elements of x larger than 1
[1] 2 3
> x[x == 4] # those elements of x that are equal to 4 (none)
numeric(0) # yields a numeric object (vector) with no entry
> x[-2] # exclude certain entries (the second one)
[1] 1 3
> x[-c(2, 3)] # if you want to exclude entries 2 and 3
[1] 1
```

Since we have not made any assignments (<-) of the operations earlier, none of them 'sticks' and the original x and y vectors are still in place.

Subsetting data frames with [,] Row selection before the comma, column selection after the comma, e.g.: d1[rows, cols]

Datasets are called 'data frames' in R (see also Box 1.4. 'Objects in R'), according to the function `data.frame`. When you read data into R (see Box 1.2. 'Reading data into R'), data frames are mostly created automatically, but as we have seen, you can also create data frames within R. Data frames can be subset in the same way, using square brackets (see the following code), but because they are two-dimensional, you have to use a comma inside the square brackets, the selection before the comma refers to rows, the selection after the comma to the columns. Individual columns of data frames can be accessed using the dollar sign ('$'), which you may think of as a hook to pull out variables (i.e. columns).

```
> x <- c(1, 2, 3)
> y <- c(2, 2, 2)
> # Create a data frame
> d1 <- data.frame(variable1 = x, variable2 = y) # with x and y as defined
earlier
> d1
  variable1 variable2
1         1         2
2         2         2
3         3         2
> names(d1) # display the header names
```

```
[1] "variable1" "variable2"
> d1$variable1 # the dollar sign allows access to variables within data frame
[1] 1 2 3
> d1$variable2[2] # access the second element of variable 2
[1] 2
> # which is equivalent to:
> d1[2, 2] # second row, second column
[1] 2
```

Note that the dollar sign can also be used in other occasions to access content of objects (e.g. to extract a *P*-value from a *t*-test, see Chapter 5 to two sample tests). The function `subset` can be used instead of the square brackets:

```
> subset(d1, variable1 > 2, select = "variable2")
  variable2
3         2
```

If you would like to add or permanently remove a variable from your dataset, you can do that using '$' and the indexing techniques shown earlier:

```
> d1$variable3 <- c(7, 8, 9) # adding a variable called 'variable3'
> d1
  variable1 variable2 variable3
1         1         2         7
2         2         2         8
3         3         2         9
> d1 <- subset(d1, select = c("variable1", "variable3"))
> d1
  variable1 variable3
1         1         7
2         2         8
3         3         9
> # or, the same can be achieved with
> d1 <- d1[, c(1, 3)]
# we can also use R's NULL object to delete a column
> d1$variable3 <- NULL
```

Here are another few functions you can use to identify, select, filter, characterise, and reshape data. We are still using x and y from earlier:

```
> which(x == 2) # useful to identify outliers, it indicates the position of
all values equal to 2
[1] 2
> unique(x) # how many unique values does a variable have?
[1] 1 2 3
> unique(y)
[1] 2
> length(x) # how long is a variable?
[1] 3
> nrow(d1) # how long is a data frame (no. of rows)?
[1] 3
> ncol(d1) # how wide is a data frame (no. of cols)?
[1] 2
> dim(d1) # d1 has three rows and two columns
[1] 2 3
> cbind(x, y) # bind columns together
     x y
[1,] 1 2
```

```
[2,] 2 2
[3,] 3 2
> rbind(x, y) # bind rows together
  [,1] [,2] [,3]
x    1    2    3
y    2    2    2
```

Sorting and ranking data makes use of `rank`, `sort`, and `order`. Closely study the following examples to figure out which function achieves what. To turn a variable around (last entry first), use `rev`.

```
> rank(c(4, 2, 3)) # indicates rank of the values (lowest to highest)
[1] 3 1 2
> sort(c(4, 2, 3)) # sorts from lowest to highest entry
[1] 2 3 4
> order(c(4, 2, 3)) # gives the index from lowest to highest entry
[1] 2 3 1
> d2 <- data.frame(variable1 = x, variable2 = y, variable3 = c(4, 2, 1) #
  with x and y from earlier
> d2
  variable1 variable2 variable3
1         1         2         4
2         2         2         2
3         3         2         3
> d2[order(d2$variable3), ] # reorder the dataset according to variable3
  variable1 variable2 variable3
2         2         2         2
3         3         2         3
1         1         2         4
```

The `sort` command only works for vectors, not for entire data frames! If you apply `sort` on a data frame column and replace the original column with the newly sorted one, it messes up the data frame because the remaining columns keep their original order. You will need practice using these functions, and you might argue that it is easier to rearrange things by hand in a spreadsheet. This may be true for small datasets, but once you start working with larger datasets, it pays off to do this in R, as not only will you be faster, but also the process is reproducible. That means that you can later come back to your code and you see precisely how the data were processed. This is not the case if you change things around manually in a spreadsheet. Although the syntax of R is relatively intuitive compared to more dedicated programming languages, it may be confusing at first how brackets, mathematical symbols, and special characters are used. Box 1.3 gives you an overview of the use of special characters in R.

Box 1.3 The use of special characters and symbols in R (Note that not all special characters listed here will be used in this introduction)

	Character	Usage
()	Parentheses	To wrap arguments of a function and in loops (e.g. for, while and repeat, see Chapter 6.2)
[]	Square brackets	For indexing of vectors, data frames, lists, and arrays

	Character	Usage
{}	Braces	In loops to delimit the contents of the loop and in if statements or functions to delimit the algorithms
“” ‘’	Quotation marks	For defining character objects (the use of double and single quotes is equivalent)
,	Commas	For separating arguments within function calls
~	Tilde	For declaring variable dependence (statistics and plots), reads: ‘as a function of’
:	Colon	To create ascending/descending integer vectors (e.g. 1:3 is the same as c(1, 2, 3) and for the specification of interactions (statistics), can be used in character strings (not recommended)
;	Semicolon	for separating individual R statements on the same line
.	Period	For model updating (statistics), can be used in character strings
$	Dollar sign	For accessing a lower level of an object, e.g. a variable within a data frame or a p-value of a model output
!	Exclamation mark	Logical negation (e.g. 5! = 4 will yield TRUE)
?	Question mark	Activates the help menu for a function
&	Ampersand	Logical ‘and’
=	Equality	To assign a value to a variable, between argument names and their values inside a function
==	Double equality	Logical comparison (‘is equal to?’) evaluating to TRUE or FALSE
<-	Assignment	To assign a value to an object, can be used in both directions (->), = can be used instead of <-
|	Pipe operator	Logical ‘or’. Also used for model specification
#	Hashtag	For comments in the code (R ignores what comes after)
%	Percent	Part of the pipe operator in *dplyr* (%>%), also for matrix multiplications (%*%) and logical operand matching (%in%)
\	Backslash	To invoke a new line (\n), tab (\t), carriage return (\r) or even an acoustic alert (\a)
/	Forward slash	For paths or as a math symbol
+	Plus sign	Apart from its normal use, R issues a ‘+’ on the console if a command is incomplete, e.g. if you have not closed a bracket
+, -, *, /, ^	Mathematical symbols	All mathematical symbols can be used in a standard way

1.5.4 Aggregating Data

The process of condensing data into a summary format is called data aggregation. R has a powerful brute-force `summary` function that gives a six-number summary of continuous variables and counts the number of observations within each level of categorical variables (see following example).

We will use the built-in dataset ToothGrowth, which contains the results of a study on the effect of vitamin C on tooth growth in guinea pigs. The data frame consists of the three variables (three columns): the response variable tooth length (len) and the two explanatory variables vitamin supplement type (supp) and the concentration of the administered vitamin C (dose). The study comprised 60 animals, each of which represents an observation resulting in 60 values (rows) in total. We will explore the data using the str (short for 'structure') and summary functions to check whether all observations are present and whether the variables are correctly coded (continuous variables should be numeric and categorical ones should be of class factor).

```
> str(ToothGrowth)
'data.frame':       60 obs. of  3 variables:
 $ len :     num  4.2 11.5 7.3 5.8 6.4 10 11.2 11.2 5.2 7 ...
 $ supp: Factor w/ 2 levels "OJ","VC": 2 2 2 2 2 2 2 2 2 2 ...
 $ dose: num  0.5 0.5 0.5 0.5 0.5 0.5 0.5 0.5 0.5 0.5 ...
```

The str output tells us that the dimensions of the data are correct (60 observations of 3 variables) and lists the variables and their type (num = numeric, i.e. continuous, Factor = categorical, see Box 1.4. 'Objects in R') along with a preview of the first few values. We can see that the response len is a continuous variable, supp is a categorical predictor variable with two levels (OJ = orange juice and VC = vitamin C powder), and dose is a continuous predictor giving the concentration of vitamin C.

From the summary output, we gather that all teeth have a positive length and the range of values makes sense. Both the orange juice (OJ) group and the vitamin C powder (VC) group have 30 observations each, so it all looks nice and balanced. The information about the narrow range of doses is not terribly useful. In fact, we will convert this continuous predictor into a factor with three levels (0.5, 1, 2, mg day^{-1}) because just three unique values would make a poor continuous predictor in any type of regression model.

```
> summary(ToothGrowth)
      len          supp         dose
 Min.   : 4.20   OJ:30   Min.   :0.500
 1st Qu.:13.07   VC:30   1st Qu.:0.500
 Median :19.25           Median :1.000
 Mean   :18.81           Mean   :1.167
 3rd Qu.:25.27           3rd Qu.:2.000
 Max.   :33.90           Max.   :2.000
```

The general summary of the data we get with the summary command is a fantastic sanity check. However, often we have groups in our data and are more interested in the group means and their standard errors making this one of the most common data aggregation routines. To demonstrate this workflow, we will use the popular *dplyr* package, which is part of a collection of packages called *tidyverse* geared towards facilitating data handling. The workflow streamlining relies on the so-called 'pipe' operator, which passes the results of one function on to the next function. The keyboard shortcut for the pipe operator is **Ctrl + Shift + M** (Windows) or **Cmd + Shift + M** (Mac). Now let us have a look at this procedure using the group_by and summarise functions, and finally aggregate the data by computing the mean and its standard error (standard deviation / $\sqrt{n}$).

```
> library(dplyr)
> tg <- group_by(.data = ToothGrowth, supp, dose) %>%
  summarise(length = mean(len,), se = sd(len)/sqrt(n()))
> tg
# A tibble: 6 × 4
# Groups:   supp [2]
  supp  dose length    se
  <fct> <fct>  <dbl> <dbl>
1 OJ     0.5   13.2  1.41
2 OJ     1     22.7  1.24
3 OJ     2     26.1  0.840
4 VC     0.5  7.98 0.869
5 VC     1     16.8  0.795
6 VC     2     26.1  1.52
```

Behind the scenes, the `group_by` function chops our data frame into subsets according to the provided grouping variables. Within the `summarise` function, we can specify as many functions to summarise our data as we want. The output is a so-called *tibble*, a new take on the good old data frame, and this new format already includes some information from the `str` command: the dimensions of the dataset, the number of levels of the first grouping variable (here `supp`) and the type of variable. Here, *fct* denotes a *factor* and continuous variables are indicated by *dbl*, which stands for *double precision* (a floating-point number format in computers), which may be the technically correct term, but it sounds very weird to the ears of the average user. If you prefer a plain data frame, then adding another pipe operator at the end, followed by the `as.data.frame()` function (with empty brackets) will do the trick.

1.5.5 Working with Time and Strings

This is an important skill we need for our basic toolbox. Text often occurs in our data files and so does temporal information (time, day, and year). Particularly handling dates/time as a variable is difficult, as they do not follow the decimal system that we are used to. There are time formats in R called POSIXt/POSIXlt/POSIXct. The detailed description of their use is beyond the scope of this introduction, and we will only give some examples that should get you going. More details can be found in the help files and in the cited literature. You can convert any time format (e.g. 'Wed 10.Mar 2008 17:00') into the R time format. Some graphics functions can handle variables of class POSIXt and format date/time information rather sensibly. For the following examples, note that they depend on the language of your installation of R.

```
> start_time <- strptime("2023/24/01 08h45", format = "%Y/%d/%m %Hh%M")
> start_time # check ?strptime for help with the coding
[1] "2023-01-24 08:45:00"
> end_time <- strptime("Mon 24.Jan 2023 17:00", format = "%a %d.%b %Y %H:%M")
> end_time
[1] "2023-01-24 17:00:00"
> end_time - start_time # calculate a time difference
Time difference of 8.25 hours
```

From personal experience, quite often it is enough to convert time variables to day of year ('doy') and day fraction ('dayfrac' spanning from 0 to 1), particularly in ecological applications. Pasted together, they yield a 3-digit number which, if you allow 5 decimals, can identify every minute of a year! When labelling graphics, you may want to reconvert the date format to a more readable one. Here are a few useful ways for converting time formats:

```
> start_time$min # note the use of $ to access a lower level
[1] 45
> start_time$hour
[1] 8
> start_time$year
[1] 123 # years since 1900
> start_time$yday
[1] 23 # day of year (from 0-365!)
> (start_time$min*60+start_time$hour*3600)/(24*3600)
[1] 0.3645833 # the fraction of the day
```

For more specialised applications (e.g. detecting past weekends and many more), also see the packages *zoo, lubridate,* and *chron,* which have to be downloaded separately.

Apart from temporal variables, you may find discrete variables whose levels are encoded in text (e.g. “treated”, “control”). Often, you want to rename those, identify them, modify them, or paste them together. Here are some useful functions for working with strings. The function `paste` does what it says, it pastes together whatever you tell it to, separated by whatever you would like (a period in the following example). If you do not want any symbol or space separating the text or numbers, use `paste0` or the argument `sep = ""`:

```
> paste("file", 1:3, "txt", sep = ".")
[1] "file.1.txt" "file.2.txt" "file.3.txt"
```

Note the way this is done element-wise, such that you obtain a variable with three values. You may want to create a new variable that combines the fertiliser level with the CO_2-treatment level in a dataset:

```
> d1 <- data.frame(co2 = rep(c("Elev", "Amb"), each = 4),
  fert = rep(c("Low", "High"), 4))
> d1
   co2 fert
1 Elev  Low
2 Elev High
...
> d1$co2fert <- paste0(d1$co2, d1$fert) # add a variable to combine the
  two + treatments
> d1
   co2fert  co2fert
1 Elev  Low  ElevLow
...
```

The values for those new variables are getting a bit long, so try this:

```
> d1$co2fert <- paste0(substr(d1$co2, 1, 1), substr(d1$fert, 1, 1))
> # takes a substring, starting at 1, ending at 1
> d1
   co2 fert co2fert
1 Elev  Low      EL
2 Elev High      EH
...
```

You can see there is little (if anything) you cannot do, and with the online help options available you are pretty sure to find a solution to most problems. If things are a little harder at first, remind yourself that your work is saved as code, and next time you do something similar, things happen a lot faster, something that is not the case in Excel!

Box 1.4 R Objects

In R, everything is an object. An object can be a data frame, a matrix, a variable (of type numeric, integer, logical, factor or character), a function (such as mean), or a model output etc. Some R functions will produce specific output, depending on what class the provided argument belongs to. For example, the function `plot` (explained in detail later) will produce different results, depending on whether you feed it a 'numeric' or 'lm' object (linear model output). The functions `lm` and `rnorm` do not need to be understood at this stage, the point here is that a function will recognise the class of the provided object and produce a (sensible) output.

```
> x <- 1:10; y <- rnorm(10)
> m1 <- lm(y ~ x) # fit a linear model to random numbers
> plot(y)
> plot(m1)
```

Confusion and frustration arise when trying to make R do things with objects of the wrong class. You cannot compute a mean of a variable that R identifies as a factor:

```
> a <- as.factor(rep(c(1:3), 3))
> mean(a)
[1] NA
Warning message:
In mean.default(a) : argument is not numeric or logical: returning NA
```

Generally, you should always know what class an object you are handling belongs to. In case of doubt, use `class` to check:

```
> class(a) # as defined above
[1] "factor"
> class("Hans")
[1] "character"
> class(c(1:3))
[1] "integer"
> class(mean)
[1] "function"
> class(sqrt(c(1:3)))
[1] "numeric"
> class(rbind(x, y))
[1] "matrix"
> class(model1)
[1] "lm"
```

The most important classes are summarised below. There are many more, but these are the most important ones. Note that you can (and sometimes must) convert objects to a specific format before analysis, e.g. if a variable is of class 'ts' (time series), but you would like to draw a simple barplot, you need to use `as.numeric` to change the class from 'ts' to 'numeric'.

(Continued)

Box 1.4 (Continued)

Object class	Description
numeric	A single decimal number or a vector of decimal numbers
integer	A single integer number or a vector of integers
character	A single character string (letters, numbers, special characters, or combinations thereof) or a vector of character strings
factor	A vector with entries corresponding to *n* levels. Useful for statistical analyses and plots
logical	TRUE, FALSE, or NA (not available, entry missing, can be numeric)
ts (time series)	A time series object facilitates the analysis of data collected over time
matrix	An $n \times m$-dimensional table with a header row. The values contained are usually of class numeric, integer, character, factor, or logical
list	Incomplete but intuitive definition: a drawer chest with *n* drawers. Each drawer can contain any of the objects listed here
lm	The output of a simple linear model from which you can extract further details, e.g. if m1 is an object of class 'lm', you can retrieve the fitted values using 'm1$fitted.values'
function	Either a built-in R function (e.g. mean) or a user-defined function
NULL	An object that exists but is not defined. E.g. you can define 'a = NULL'

With as.numeric or as.character, one can (try to) change the class of an object (coerce an object), which cannot always be achieved. An R object may contain more information than displayed by simply typing its name. A useful way to access parts or additional information of an object is:

```
> names(m1) # using m1 from earlier
 [1] "coefficients" "residuals" "effects" "rank" "fitted.values"
 [6] "assign" "qr" "df.residual" "call" "terms" "model"
```

The class 'factor' merits special attention, as it often leads to confusion. Factors memorise their levels, even if those are no longer present, which can cause trouble in a plot. If you have turned a variable into a factor, arithmetic operations with it are no longer possible (often pointless anyway). Sometimes, we have factors with actual numerical meaning, that we wish to make numeric. These factors must first be turned into a character to ensure correct conversion into a numerical variable:

```
> fac <- as.factor(c(25, 50, 100))
> fac
[1] 25  50  100
Levels: 25 50 100
> as.numeric(fac) # wrong, just returns the intrinsic levels
[1] 1 2 3
> as.numeric(as.character(fac)) # correct
[1]  25  50 100
```

2

Measuring Variation

2.1 What Is Variation?

If you measure your body height 10 times on a single day, you will not measure the exact same value every time. If you measure the circumference of all heads of newborns on a given day at a hospital, you will receive a different value for every baby's head you measure. Even if you weigh plastic parts that are mass produced by a 3D printer, you will get a range of values if your balance is accurate enough. Variation is so inherent in both the natural and the human-made world that a deep understanding of its nature, its causes, and the possibilities to describe it, are fundamental. In fact, *assigning* variation to certain causal drivers can be viewed as *the core purpose* of statistical analyses. In this chapter, we will look at how variation is partitioned, the two different types of variation, explained vs. unexplained variation, and ways of describing variation both visually and numerically. We will rely on the basic R skills acquired in Chapter 1.

2.2 Treatment vs. Control

Any statistical analysis is ultimately *comparative*. The (scientific) statement 'This treatment helps prevent arthritis' actually means 'On average, the odds of getting arthritis for people receiving this treatment are lower *compared to* people in a control group who do not receive the treatment'. The second (comparative) part of the sentence is often not expressed explicitly, yet extremely important. Without it, we find ourselves in a 'no-control' scenario, or, in other words, we assign variation to a cause (the drug) without knowledge of any pre-existing, or background variation or concurrent changes. As mentioned earlier, humans are by intuition deeply 'unstatistical' in their thinking, so our daily lives and conversations are filled with 'no-control' statements and perceptions. You will often hear statements like 'Since I have started eating this type of food I sleep much better', or 'this summer was very windy'. In both cases, however, there is a lack of a control, i.e. we do not know if the person would have slept better *anyway*, even without eating a particular type of food, and we do not know if the summer described as 'windy' was actually windier than the average summer at this location. While we cannot expect anyone to back up simple statements like those above with control data, we need to do this diligently in the scientific context. Although the word 'treatment' makes us think of a drug or a medical procedure, in the statistical environment, it is used to describe any deliberately imposed change to an experimental system. In plant

'No control' statements are common in our everyday language, but must be avoided in statistical reasoning!

R-ticulate: A Beginner's Guide to Data Analysis for Natural Scientists, First Edition.
Martin Bader and Sebastian Leuzinger.

Companion website: www.wiley.com/go/Bader

science, this could be a fertiliser treatment, in food science, it could mean the use of a new ingredient, and the use of a novel compound in engineering. The ultimate goal is always to disentangle the effect (or variation) caused by a treatment from the variation that is present in a system in the absence of that treatment. The variation is measured in the response variable (see Section 1.4.3 on Predictor and Response Variables in Chapter 1). In the simplest case, the implementation of this approach usually works via dividing your study objects (e.g. patients, plants) into two groups – one receiving the treatment (the treatment group), and one not receiving it (the control group). If you would like to test the effect of a certain drug on the heart rate or length of mice, you will have a certain number of mice receiving the drug, and, concurrently, an equal number of mice not receiving the drug. This allows you to partition the observed variation into variation that is due to the treatment and variation that is present regardless of the treatment.

The terms 'treatment' and 'control' are generic and not only used in a medical context.

2.3 Systematic and Unsystematic Variation

Think of a darts player who is told to hit the bull's eye (the red centre of the target) 10 times in a row. While they might hit the bull's eye a number of times, they will likely miss it by some distance on some occasions. The direction and distance from the target will likely be random, i.e. they might be aiming a little bit too high in some cases, but equally often too low, too far right, too far left, and so on. Therefore, the deviation (or variation) from the bull's eye is random, or unsystematic (left side of Figure 2.1). You could also view this as *noise* that has been added to a perfect hit of the target. On the right-hand side of Figure 2.1, all darts seem to have gone off-target in the same manner (they were biased towards the top right corner). Here, the deviation (or variation) is not random, but *systematic*.

Let us now think about the underlying causes for the two types of variation. The reason for the unsystematic variation observed on the left in Figure 2.1 is pure randomness and not attributable to a specific cause. The systematic variation shown on the right in Figure 2.1 is

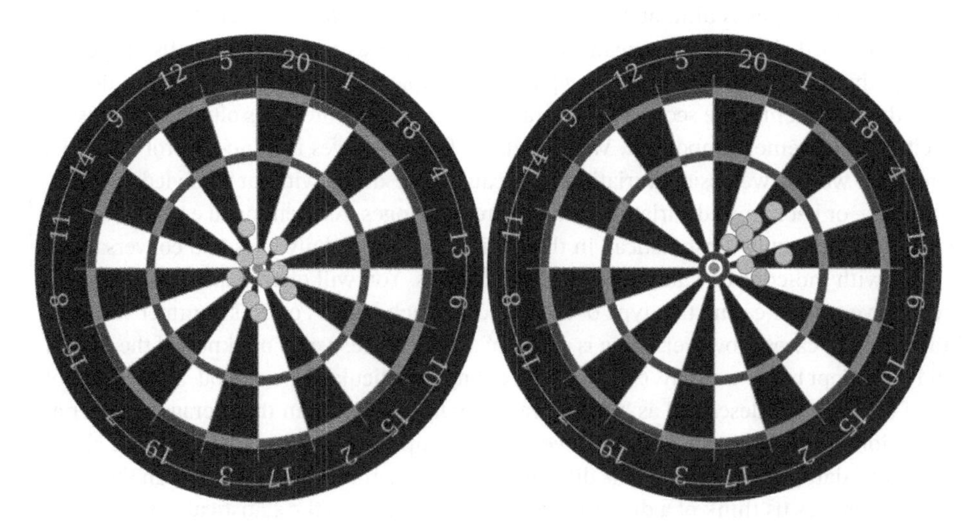

Figure 2.1 An illustration of unsystematic and systematic variation. On the left, the darts player deviates randomly (unsystematically) from the bull's eye (unsystematic variation), and on the right side, the deviation is systematic (systematic variation).

Table 2.1 Possible sources of unsystematic and systematic variation in a simple experiment with mice, see text.

Stage of experiment	Possible source of unsystematic variation	Possible source of systematic variation
Set up of the experiment	The mice come from a more heterogeneous source (e.g. they are not all of the same age, or genetically more diverse)	The mice happen to be unusually small/large
	The mice are kept under more heterogeneous environments (exposed to varying light, temperature)	The mice are by accident exposed to more or less light/heat
Administration of the treatment	The person who prepares the injections is tired and makes random mistakes while pipetting	The pipette to prepare the injections is biased, e.g. the volume is too large or too small
	Several different experimenters perform the injections	One experimenter takes care of the treatment mice, a different one of the control mice
Measurement of the response variable	The calipers to measure the mice are old and difficult to read	The calipers used are biased (they systematically indicate values too low or too high)
	The measurements are not done carefully, random mistakes are introduced	The measurements of the treatment/control groups are carried out by different experimenters

The mentioned sources of variation can apply to the treatment and/or control groups. Most of these sources of variation can be avoided or minimised if we are aware of them.

different – it is predictable and is added to the random variation. The cause for this type of variation could be that the darts are slightly biased, or the darts player had a recent injury that makes them aim systematically too far to the top right. In short, systematic variation means bias, unsystematic variation noise. Put simply, we want to either avoid both *or* be able to understand the origin of systematic variation, both ultimately blur the view on our results or 'signal'.

The goal is to link systematic variation to a causal driver (a predictor variable)

To train ourselves to become aware of all the occasions where these two types of variation can sneak in, we need to look at examples. Imagine an experiment where you treat 10 mice by injecting them with a drug, and you keep 10 mice as a control group. Note that the best practice is to also inject the control mice, but with a neutral substance that does not affect the mice. This is to make sure that the observed response is not simply due to the effect of the (physical) injection, for instance because the animals become stressed (see Chapter 4 on confounding effects). Let the response variable (the variable of interest, see Section 1.4.3) be the size of the mice, say the length of the body. Table 2.1 lists possible sources of unsystematic and systematic variation for this simple experiment.

2.4 The Signal-to-Noise Ratio

Understanding what proportion of the variation in our data is 'signal' (caused by an effect) and what proportion is 'noise' (random variation) presents the most fundamental challenge

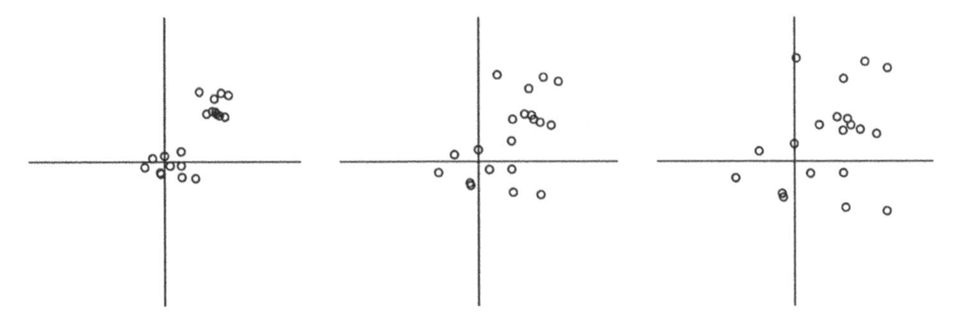

Figure 2.2 The signal-to-noise ratio visualised. Two groups of data can be made out on the left despite the noise. In the middle, the noise is increasing and the two groups can barely be identified (despite using the same mean on both axes). On the right, still using the same mean values along the two axes, the noise is so large that the two groups can no longer be distinguished, although the difference in means still exists.

in statistics. While at times there might not be a signal (or perhaps there is one, but it is only as strong as the underlying noise), we are certain to always see noise in our data. Defining a signal-to-noise ratio is therefore extremely useful. If the signal to noise ratio takes on a value of 1, the signal is entirely drowning in noise. The larger the ratio, the clearer the signal. A useful analogy is the communication via a VHF radio (VHF = very high frequency). Depending on the signal strength, you may hear the person at the other end loud and clear on top of a small carpet of noise. As the connection deteriorates, you will understand less and less of what the other person says (as the signal-to-noise ratio decreases), until the voice cannot be understood at all (at a signal-to-noise ratio of 1). Figure 2.2 illustrates this graphically. The mean values of the two groups that are visible on the left remain unchanged (the signal in this example), but the noise increases from left to right, until the signal becomes completely obscured by noise. The concept of the signal-to-noise ratio is of such fundamental importance, that we will come back to this concept in Chapter 3.

2.5 Measuring Variation Graphically

Figure 2.2 shows variation symbolically in a two-dimensional space, much like in the darts example above. If we want to visualise the variation in a single continuous variable, we have two common options: a simple 'dot' chart, or a histogram. The histogram can be viewed as the most basic, yet the most important plot in statistics. The idea is to set up 10 to 20 categories or 'bins' to which the values are assigned. For example, if our variable is 'age', you could set up 10-year bins, so 0–9 years, 10–19 years, etc., up to 90–99 years. Seeing how 'full' each of those 10 bins will be once all values are assigned will give us an idea of the variation present in our variable. For instance, if only the bins 0–9 and 10–19 years fill up, we know that the variation in the variable 'age' is low, and most people are teenagers or even younger. A histogram also tells us about outliers, and about the variable's *distribution* (see Chapter 3). All too often, people proceed to data analyses without having looked at histograms of their variables. The 'dot' chart is another elementary plot, in which the values of a variable are plotted on the x-axis one by one against their index, i.e. the first value is plotted on the bottom of the graph, then the second value on top, etc., with the last value at the top of the graph. Figure 2.3 gives

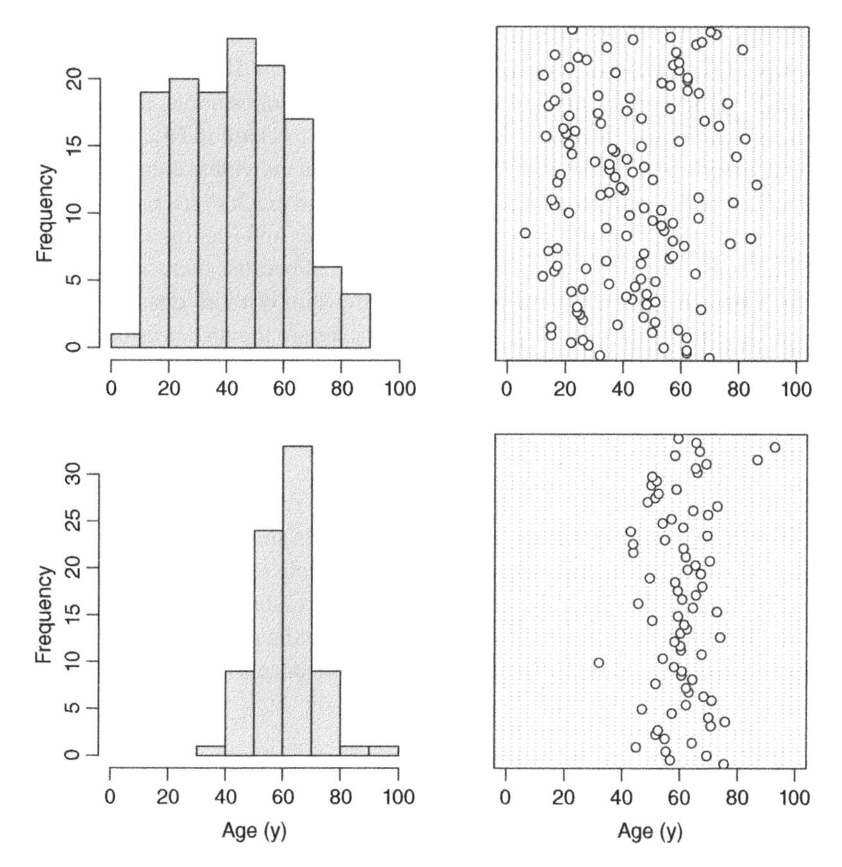

Figure 2.3 The top two plots show the variation in age of 230 members of a table tennis club in Finland, while the bottom two show the variation in age of the 78 members of a bridge club in Northern England. The left-hand side uses histograms, the right-hand side dot charts, where the y-axis represents the position of the values inside the variable (from 1 to the 230 or 78, respectively).

an example of both types of plots, showing the variation in age of members of a table tennis and a bridge club (bridge is an English card game). While the histogram shows the wider spread of age in the tennis table club at a glance, the dot chart is useful to draw attention to the outliers (the 32 years old and the two over 90 years old members of the bridge club). The code that was used to produce the plots in Figure 2.3 is

```
> ## x contains the ages of the tennis table club members:
> hist(x, xlim = c(0, 100))
> dotchart(x, xlim = c(0, 100))
> ## y contains the ages of the bridge club members:
> hist(y, xlim = c(0, 100))
> dotchart(x, xlim = c(0, 100))
```

A histogram shows the variation or spread of a single variable.

2.6 Measuring Variation Using Metrics

Graphically assessing variation is great if we need to get a quick idea of the nature of our data. If we want to quantitatively compare variation, and in order to separate and assign variation in models, we need numbers. A first intuitive step to describe the variation in a variable is to look at the distance of the values in a variable relative to the mean. If the distances are large, the variation is large, if they are small, the variation is small. For example, if you look at Figure 2.3, the mean age of the table tennis club members looks to sit around 45, while the mean of the bridge member club is around 60. The distances of the individual data points to the mean are much longer in the example of the table tennis club and much shorter in the case of the bridge club. So we could add up those distances in both cases and use this as a metric for the variation in the two variables. The problem with this is that because there are exactly as many negative numbers as there are positive numbers, the sum of those distances will always be zero – not helpful. To alleviate this, we can simply square the distances between the mean and the data points, as this eliminates negative values. The metric we have now arrived at is the *sum of squares* or *SS*:

$$\text{Sum of squares} = \sum (x_i - \bar{x})^2$$

with $\bar{x}$ representing the mean, and x_i the individual values or observations. However, to use this metric to compare the variation between different variables is still not useful, because as evident from the example of the club members above, the table tennis club has many more members, and will therefore yield a higher *SS* value for that reason alone. We therefore need to bring the number of values contained in the variable into the game. This yields the variance, a standardised form of the sum of squares, often abbreviated as s^2 (it will become clear why):

$$\text{Variance} = \frac{\sum (x_i - \bar{x})^2}{n-1} = s^2$$

In R, we can easily compute the sum of squares manually and then derive the sample size, but you will likely use the in-built function `var` to do so

```
> x <- c(3, 5, 1, 3, 7, 6, 7, 4, 2) # creating a variable
> n <- length(x) # extracting the length (n)
> mean_x <- mean(x) # calculating the mean
> ss <- sum((mean_x - x)^2) # calculating the sum of squares
> variance <- ss/(n-1) # calculating the variance
> variance
[1] 4.694444
> var(x) # using the R function
[1] 4.694444
```

Here, n is the number of values that are contained in the variable (the sample size). You will note that we have divided by one less than that exact number, and you might wonder why this is. For an exhaustive explanation, you will have to turn to more mathematically focused resources, but here is a brief intuitive explanation. Imagine you want to estimate the variation in the body height of the male students of a small university with 1230 students, using the variance as a metric. Let us assume that the tallest student is 203 cm, and the shortest 152 cm. If you are taking a small sample, say five students only, then the probability that your sample includes those most extreme values (or any other very low or high values), is understandably small. This will lead to an *underestimation* of the variability in male students' heights. However, if you were to take a sample of 1000, the chances of including some of those extreme values are very high. The division by $n-1$ corrects for exactly this, the correction is larger at low vs. high sample size. Dividing by 4 instead of 5 elevates the value notably. Contrarily, at high n, the correction is only small, as it makes little difference whether we divide by 1000 or by 999. This also explains why you can just divide by n and no longer by $n-1$ once you sample the entire population, as you will have the extremes included with certainty. The quantity $n-1$ is also termed the 'degrees of freedom'. Sloppily expressed, you can say that the calculation of the mean, which you used to compute the variance 'cost' you one degree of freedom. This can be understood when you think of someone calculating the mean (or ultimately the variance) of say five numbers: once you have calculated that mean, you can freely choose four of those numbers, but not the fifth, as the fifth number is locked in to match the mean which you have already calculated!

Since the variance squares the values and thus also the units of our variable, we cannot directly compare it to the mean. For example, if our variable is 'weight' (in kg), then the calculated variance would be in kg^2. Would it not be nice to have a metric that uses the same units as the original variable and the mean? We therefore also define a *standard deviation* as the square root of the variance:

The standard deviation is the square root of the variance.

$$\text{Standard deviation} = \sqrt{\frac{\sum (x_i - \bar{x})^2}{n-1}} = s$$

Both the variance and the standard deviation are important metrics to remember, the variance is used more often in statistical models, while the standard deviation is referred to when we want to characterise the variation (or spread) in a variable. In R, the standard deviation is readily derived from the variance, or computed directly using the function `sd`:

```
> sqrt(var(x)) # using the x from above
[1] 2.166667
> sd(x)
[1] 2.166667
```

2.7 The Standard Error

Imagine you are tasked to estimate the mean and the variation in the weight of a type of fish in a lake. You might choose to use 3, 10, 100, 300, 500, or even 700 samples to do so. As you increase your sample size, your estimate for the mean will obviously improve. Similarly, your estimate for the variation in size (the standard deviation) will also improve. For the standard deviation, this is shown in the left panel of Figure 2.4 – the estimate improves and approaches its true value of 1 (we only know the true value because this is a simulation). In other words, you have become better at answering the question: 'What is the mean and variation of the variable "fish weight?"'. What you cannot make a statement on, but what is also of interest, is *how well* you could estimate the mean. It is therefore useful to come up with a metric for the variation in a variable that reflects exactly that, i.e. decreases with the sample size (right panel in Figure 2.4). We use the standard error for this:

$$\text{standard error} = \frac{s}{\sqrt{n}} = se$$

Mathematically, the standard error is the standard deviation of many sample means, characterising the variance of those means. It is most important to remember the two fundamentally different questions that the two metrics answer: the standard deviation characterises the

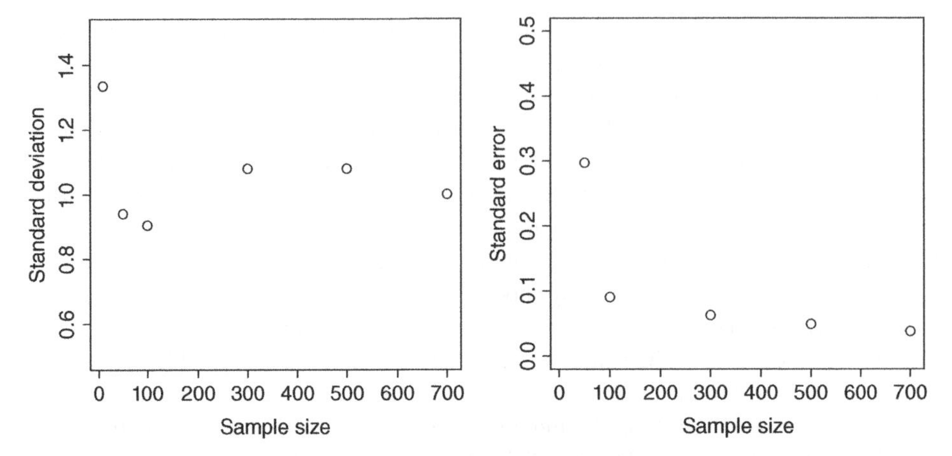

Figure 2.4 A visual representation of the standard deviation versus the standard error. The two metrics show two different things: the standard deviation on the left shows the variation in a variable. As the sample size increases, the estimate becomes more precise (the true standard deviation is equal to 1). The standard error on the right shows how well we are able to estimate the mean; it logically decreases with the sample size.

variation or spread in a variable, while the standard error indicates how well we are able to estimate its mean. With increasing sample size, the standard deviation becomes more accurate, and the standard error decreases (as we gain confidence in estimating the mean).

In R, there is no in-built function to compute the standard error, so we simply divide the standard deviation by the square root of the sample size:

```
> sd(x)/sqrt(length(x)) # still using the x defined above
[1] 0.7222222
```

For an elegant solution to compute the standard error in R, see Box 2.1 on writing functions!

Box 2.1 Writing functions in R

The anatomy of a function is very similar to cooking with a recipe – see the schematic below: the function name is like the name of the dish, the ingredients are the function arguments, and the directions in a recipe correspond to the function algorithm (see figure below). Writing a function in R is in essence creating an object of class 'function', which then works much the same as existing functions in R such as `data.frame` for example. In fact, if you type the name of functions into the console (try 'data.frame' without parenthesis and arguments), R will print the code of the function as it was programmed in R. The basic syntax will always look like this:

```
> Myfunction <- function(x) {algorithm of the function}
```

The process of writing a function is best explained using an example. As there is no built-in function to compute the standard error, let us programme one. As per Section 2.7, what we want the algorithm to do is to compute the standard deviation of a variable, and then divide it by the square root of the number of values contained in that variable. This will look a bit like this, assuming *x* is our variable of interest:

```
> sd(x)/sqrt(length(x))
```

We now place this algorithm into the parentheses of the structure outlined above, calling the function `se`:

```
> se <- function(x) {sd(x)/sqrt(length(x))}
```

The function `se` can from now on be used, e.g. by typing

```
> se(c(3, 5, 7, 4, 8))
[1] 0.9273618
```

It is now easy to see how this simple function to compute the standard error can be elaborated on. For example, we can include a warning in case you were to feed the function a non-numeric argument:

```
> se <- function(x) {
  if(is.numeric(x)) {
    sd(x)/sqrt(length(x))} else
    {print('Argument non-numeric')}
  }
```

Finally, let us test our home made function:

```
> se('John')
[1] "Argument non-numeric"
```

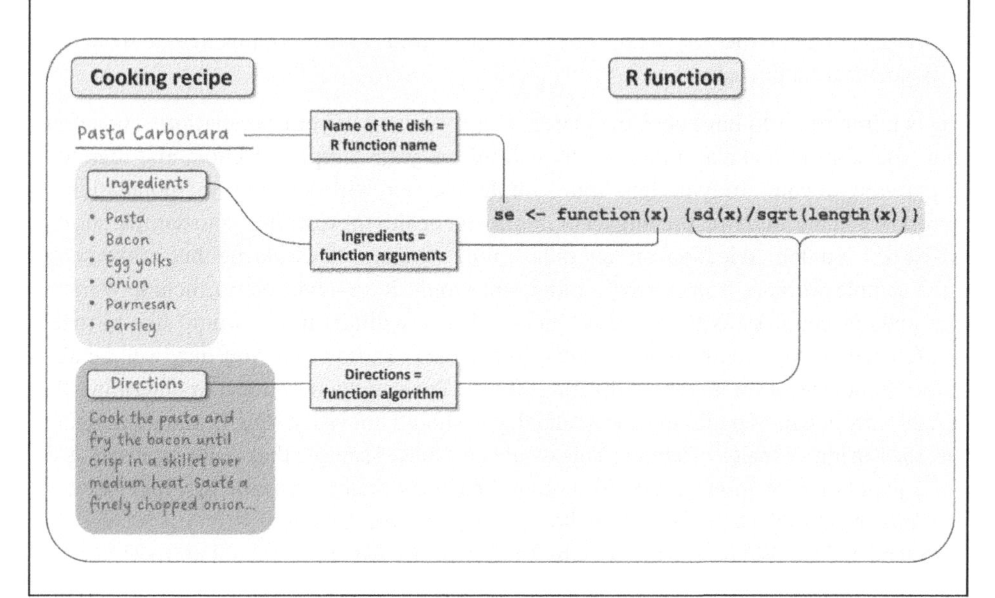

2.8 Population vs. Sample

A 'population' in a statistical sense refers to the entirety of subjects that are under investigation, it does not just refer to 'people'. For example, if we study African elephants, we will not be dealing with *every* African elephant on the planet, but our findings will likely be applicable to all elephants, i.e. the whole 'population' of African elephants. Note that regardless of whether you study bacteria, rocks, or plastic parts – all actual or potential subjects that fall into that respective category will be referred to as the 'population'. Contrary to this, any subset of a population is called a 'sample'. In the process of planning observational studies and experiments, it is essential to know what our population of interest is. As researchers we rarely get access to the entire population, but mostly we are working with a sample or several samples taken from the population. Three examples will illustrate this:

(1) If you are studying drugs that help lung cancer patients, your sample might be small. Perhaps just 20 lung cancer patients in your local hospital gave their consent to participate in the study. Your sample therefore consists of the 20 patients you have access to. The population for which your findings are relevant, however, are likely all lung cancer patients on the planet, now and in the future, as the potential benefit of the developed drugs will be of use to all of them.
(2) Assume you are working for a provider of grass seeds, and you want to find out what the germination rate of those seeds is, once they are packaged and sold. Here, your population consists of all the grass seeds of that type that the company sells at that time. Your sample might be a collection of 50 grass seed packages from retail shops that sell the product. A trial during which you sow the grass seeds could follow, and your results will be relevant for all grass seeds produced by that company.
(3) Still working for that same company, you might one day receive a tray of seeds with the question: 'Can you work out how many of these seeds germinate?'. In this case, you are given access to the entire population, as the task is to work out the proportion that germinate of the very set of seeds presented. Your sample is equivalent to the entire population – a rare case.

Why is it important to have absolute clarity over the extent of your population? The population you want to make an inference on will inform your sample selection, and therefore the planning of your study design. Your sample selection will vary with the population of interest. If you are studying the impact of mining on health in Australia, your sample should consist of a random selection of Australian mine workers. It would neither make sense to the sample workers from a single mine, nor would it be advisable to include workers from another country (as the circumstances will likely differ) if you would like to make an inference on Australian mine workers. In example (1), however, you may well be able to draw conclusions on anyone with lung cancer. Put simply, if you are to determine the average body height of males in New Zealand, you should not visit a New Zealand basketball club, measuring all males in a team. This would be a biased sample that is not representative of the population of interest, i.e. all of New Zealand's males. Instead, you may want to randomly select subjects using a phonebook or government database.

The sample has to be selected to best represent the population you want to make an inference on.

We tend to use different symbols when referring to statistics that relate to either the population or the sample. Generally, Greek letters are used for population statistics and alphabetic letters for sample statistics. The population mean is normally referred to as μ, while the sample mean is referred to as $\bar{x}$. For the population standard deviation and variance, we use σ and σ^2, respectively, and for the same sample statistic, we use s and s^2.

In this chapter, we have gone through important notions and concepts that will serve as a foundation for the remaining chapters. We have seen where variation in datasets might originate from, and how we can partition it into either systematic or unsystematic variation. The signal-to-noise ratio is the (explained) systematic variation divided by the total unexplained variation, and we have seen that the overarching goal of statistical analyses is to maximise this ratio. We introduced a series of R skills, including the visual representation of variation, and how to write your own functions. You should also have a good overview of how the sum of squares relates to the variance and how and when to use the standard deviation versus the standard error. Finally, we made a distinction between the 'population', and a sample.

3

Distributions and Probabilities

3.1 Probability Distributions

We hear the term 'distribution' a lot in statistics, so what exactly is it? A distribution, or more specifically, a ***probability distribution***, is a mathematical function that links each value of a random variable with its probability of occurrence in the range of possible values (the so-called sample space). The ***probability*** gives the likelihood of an event or observation to occur and ranges from 0 (impossibility) to 1 (certainty). So, distributions and probabilities are inextricably linked. If we can figure out the underlying distribution of any variable, then we can calculate probabilities of certain events and make predictions for their occurrence. What might be a bit confusing in the beginning is that not only the data that we collect in our experiments follow certain distributions (at least roughly), but also other variables such as test statistics – but we will elaborate on this later.

When talking about distributions, we need to distinguish between *continuous* and *discrete* random variables (see Section 1.5.2). The distribution of a continuous random variable is characterised by a probability density curve, which is derived from an equation called the ***probability density function*** **(PDF)** (Figure 3.1). Such a probability density curve indicates the relative probability of occurrence of certain values of the random variable. To be precise, the PDF gives the probability of the random variable to fall in a particular range of values (rather than taking on a specific value).

Why can we not get the probability of a continuous random variable taking on an exact value? That is because there is no such thing as a point probability for continuous variables since there is an infinite set of possible values to start with (in fact, the absolute probability to obtain a particular value is 0). However, we can certainly define a very narrow range, allowing us to hone in on a certain value of interest. The area under the probability density curve equals 1, representing the total probability of 100% for the entire range of possible values.

Probabilities range from 0 to 1, and the equivalent percentages from 0 to 100%

By definition, discrete random variables take on specific, distinct values, and their distribution is linked to a ***probability mass function*** **(PMF)**, which gives the probability of the discrete random variable to be *exactly* equal to a certain value (Figure 3.1). Here, point probabilities exist because of the discrete nature of the variables.

The ***cumulative distribution function*** **(CDF)** gives the probability that a discrete or continuous random variable X (e.g. number of cubs in a litter as a discrete variable, body size as a continuous variable) will take on a value less/greater than or equal to x (x represents a value in the sample space, i.e. the set of all possible outcomes) (Figure 3.1). Essentially, the

R-ticulate: A Beginner's Guide to Data Analysis for Natural Scientists, First Edition.
Martin Bader and Sebastian Leuzinger.

Companion website: www.wiley.com/go/Bader

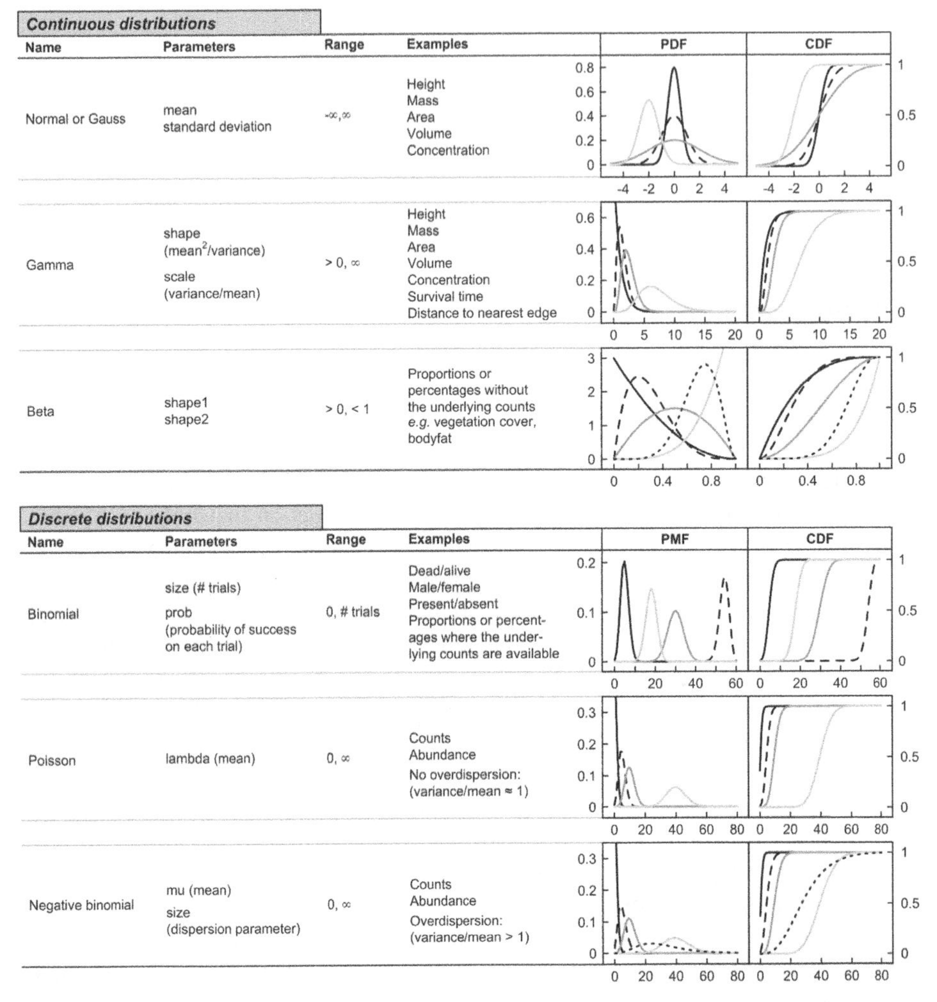

Figure 3.1 Summary of important probability distributions along with their **probability density function** (PDF) or **probability mass function** (PMF) and the **cumulative distribution function** (CDF). The probabilities given by the PDF sum up to 1 (100%) represented by the area under the curve (y-axis: probability density, x-axis: values of the random variable within the section of the sample space specified by the distribution parameters). The PDF can thus be used to determine the probability of the random variable x to fall within a particular range of values (continuous probability functions are defined for an infinite number of values over the range of possible values and thus the probability at a single point on the x-axis is always zero). The CDF gives the probability (y-axis ranging from 0 to 1) that a real-valued random variable X will take on a value less than or equal to x (the theoretical quantiles given on the x-axis). The examples were drawn using the `curve` function in combination with the density functions for the respective distribution to obtain the PDF curves (e.g. `dnorm`) and the corresponding probability functions to obtain the CDF curves (e.g. `pnorm`). The following parameters were used to create the PDF (PMF) and CDF curves in this figure: **normal distribution**: solid black = mean 0, standard deviation (sd) 0.5, dashed black = mean 0, sd 1, solid dark grey = mean 0, sd 2, solid light grey = mean −2, sd 0.75; **gamma distribution**: solid black = shape 1, rate 1, dashed black = shape 3, rate 2, solid dark grey = shape 5, rate 2, solid light grey = shape 7, rate 1; **beta distribution**: solid black = shape1 1, shape2 3, dashed black = shape1 2, shape2 5, solid dark grey = shape1 2, shape2 2, solid light grey = shape1 5, shape2 7, dotted black = shape1 7, shape2 3; **binomial distribution**: solid black = size 20, prob 0.25, dashed black = size 60, prob 0.9, solid dark grey = size 60, prob 0.5, solid light grey = size 30, prob 0.6; **poisson distribution**: solid black = lambda 1, dashed black = lambda 5, solid dark grey = lambda 10, solid light grey = lambda 40; **negative binomial distribution**: solid black = mu 1, size 20, dashed black = mu 5, size 10, solid dark grey = mu 10, size 30, solid light grey = mu 40, size 50, dotted black = mu 30, size 5.

CDF cumulatively sums up the area under the PDF (or PMF) curve from the lower to the upper boundary of the sample space.

All distributions have one or more parameters that control the shape of the probability curve. Continuous distributions often have a *location* and a *scale* parameter. The location parameter determines the position of the curve along the x-axis, whereas the scale parameter controls the spread of the curve. For example, the normal distribution has two parameters: the mean (location parameter) and the standard deviation (scale parameter).

An overview of some common probability distributions including their PDFs and CDFs in the natural sciences gives Figure 3.1.

3.2 Finding the Best Fitting Distribution for Sample Data

Now that we have a fundamental grasp on distributions, the big question is how to determine which distribution our data follow?

We have multiple options to figure this out:

- draw a ***histogram***
- draw a ***quantile-quantile plot***
- draw a ***probability-probability plot***
- perform a ***goodness-of-fit test***

3.2.1 Graphical Tools

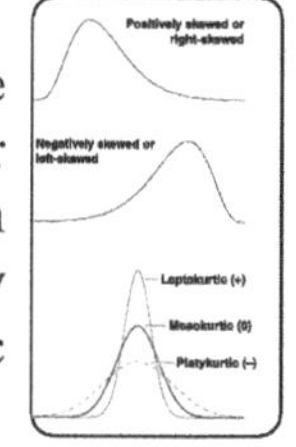

A histogram gives us a quick idea of the shape of the distribution of a variable, i.e. whether the distribution is symmetric or asymmetric (skewness), heavy- or light-tailed (kurtosis), uni- or multimodal, and so on. We can visually compare it to histograms of random variables from known distributions, but this is a rather informal approach. Adding a probability density curve of a presumed distribution to the histogram of the sample data is a more systematic method (Figure 3.2).

A quantile-quantile plot (**Q-Q plot**) is a graphical tool for comparing two probability distributions (Figure 3.2). The idea is to assess how closely the distribution of two sets of variables agrees. Instead of comparing the raw sample data to random numbers generated from the presumed distribution, this technique uses their quantiles, which are simply values that divide the distribution into equal-sized intervals (see Section 3.3). The Q-Q plot can therefore be thought of as a calibration plot, where our data (sample quantiles) are sorted in ascending order and plotted as a function of the theoretical quantiles of the presumed underlying distribution. If the points fall about a straight line, we can safely assume that our data follow the tested distribution.

In a similar vein, the probability-probability plot (**P-P plot**) compares two cumulative distribution functions to determine whether they represent the same distribution (Figure 3.2). The P-P plot is less frequently used than the Q-Q plot but is interpreted in the same way: the points should cluster around a straight line and gross deviations from it indicate departures from the specified distribution.

We can conveniently obtain all those plots using the `fitdist` function (R package *fitdistrplus*, Delignette-Muller and Dutang, 2015), which requires our data and the distribution we would like to test as inputs (Figure 3.2). The distribution needs to be specified in the form of its probability density function, indicated by a 'd' before the distribution name, e.g. `dnorm` for the probability density function of the normal distribution or `dpois` for the one of the

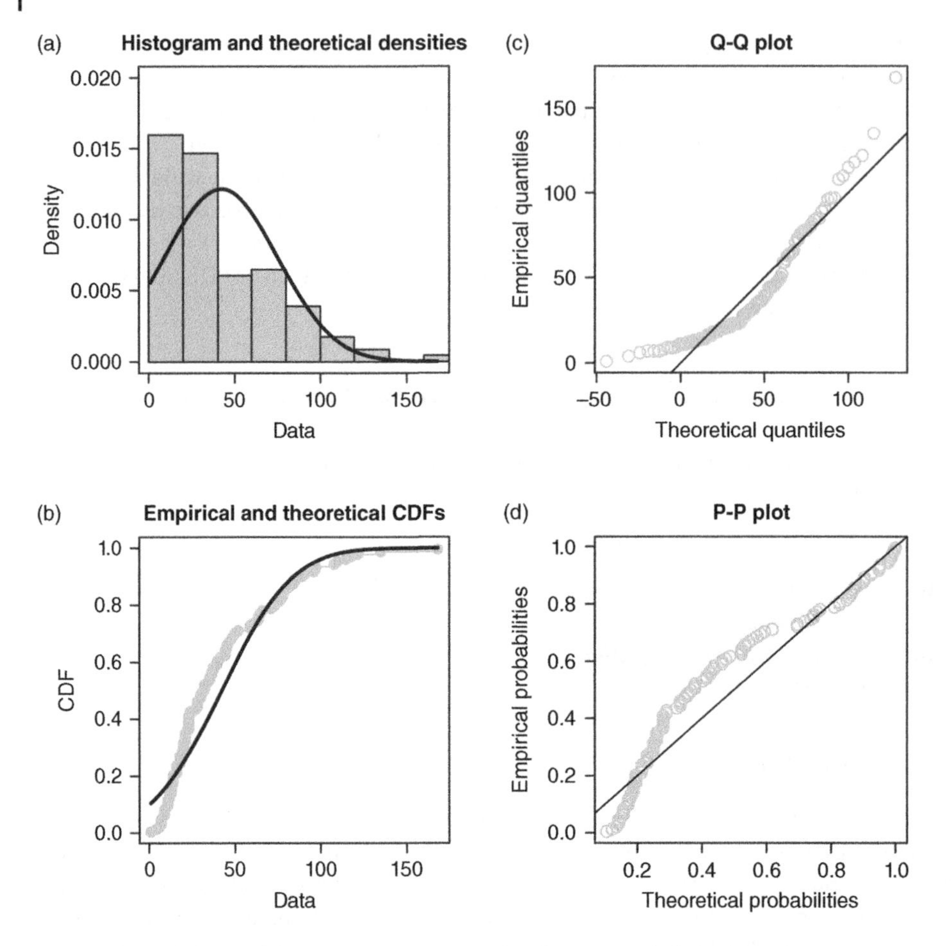

Figure 3.2 (a) Histogram of the ozone concentration from the built-in `airquality` dataset including a probability density curve assuming a normal distribution, (b) empirical (based on sample data indicated by black circles) and a theoretical cumulative distribution function (CDF, black line), here the theoretical distribution is assumed to be the normal distribution, (c) quantile-quantile plot (Q-Q plot) displaying the quantiles of our sample data (empirical quantiles) as a function of the theoretical quantiles from a normal distribution (including a straight reference line passing through the 1st and 3rd quartile), and (d) in an analogous manner a probability-probability plot (P-P plot).

Poisson distribution (type `?Distributions` in the console and hit Enter to get an overview of the distributions and related functions included in the base installation of R). We will use the `Ozone` variable in the built-in dataset `airquality` to demonstrate the procedure. The ozone concentration represents a continuous variable so that a normal distribution seems to be a natural starting point in our search for the best-matching distribution.

```
## Load required package
> library(fitdistrplus)

## Load the built-in 'airquality' dataset
> data("airquality")

## Create an NA-free ozone vector
> o3 <- as.numeric(na.omit(airquality$Ozone))
```

```
## Fit the normal distribution to the ozone data
> norm <- fitdist(data = o3, distr = dnorm)

## List with the estimated distribution parameters
norm

Fitting of the distribution 'norm' by maximum likelihood
Parameters:
     estimate Std. Error
mean 42.12931   3.049618
sd   32.84539   2.156406

## 4-panel figure with the graphical tools for distribution comparisons
> plot(norm)

## See the help file for graphcomp for individual plots, e.g.:
> qqcomp(norm)
```

Clearly, our data deviate substantially from the theoretical densities, quantiles, and probabilities of a normal distribution (Figure 3.2). The histogram shows that our ozone data are strongly right-skewed (longer tail to the right), meaning that, although most of the observations reside in the lower range, there are a few very high values.

The gamma and inverse Gaussian distributions can 'reach high up' and are well-suited for describing right-skewed continuous data.

Let us therefore repeat the previous analysis, this time testing our data against the gamma distribution.

```
## Compare the ozone data to the gamma distribution
> gamma <- fitdist(data = o3, distr = dgamma)
## Validation plots
> plot(gamma)# Figure 3.3
```

The close alignment of our data with the tested theoretical distribution in Figure 3.3 indicates that the gamma distribution provides a much better fit to our ozone data than the normal distribution. So, where does that leave us? We now know that applying a statistical model assuming normally distributed data (more specifically normally distributed model errors) would be inappropriate and lead to biased inference. We should rather use a statistical model that allows us to specify alternative distributions, such as the gamma distribution identified here, and this type of model is called a generalised linear model (GLM). GLMs can also be used in the simple case of a two-sample comparison as a more robust and appropriate alternative to the t-test or Wilcoxon test if our data follow a distribution other than the normal distribution.

3.2.2 Goodness-of-Fit Tests

Another way of testing whether our sample data come from a certain distribution is the use of goodness-of-fit tests. The ***Kolmogorov–Smirnov*** test, the ***Cramér–von Mises*** test, and the ***Anderson–Darling*** test are perhaps the most popular goodness-of-fit (GOF) tests for all sorts of distributions. The latter two tests are modified, more powerful versions of the *Kolmogorov–Smirnov* test. In addition, the ***Shapiro–Wilk*** test is widely used for normality testing. The null hypothesis of GOF tests assumes that the distribution of the sample data is not significantly different from the specified (theoretical) distribution (e.g. a normal distribution with the same mean and standard deviation as our sample data). Therefore, in GOF tests

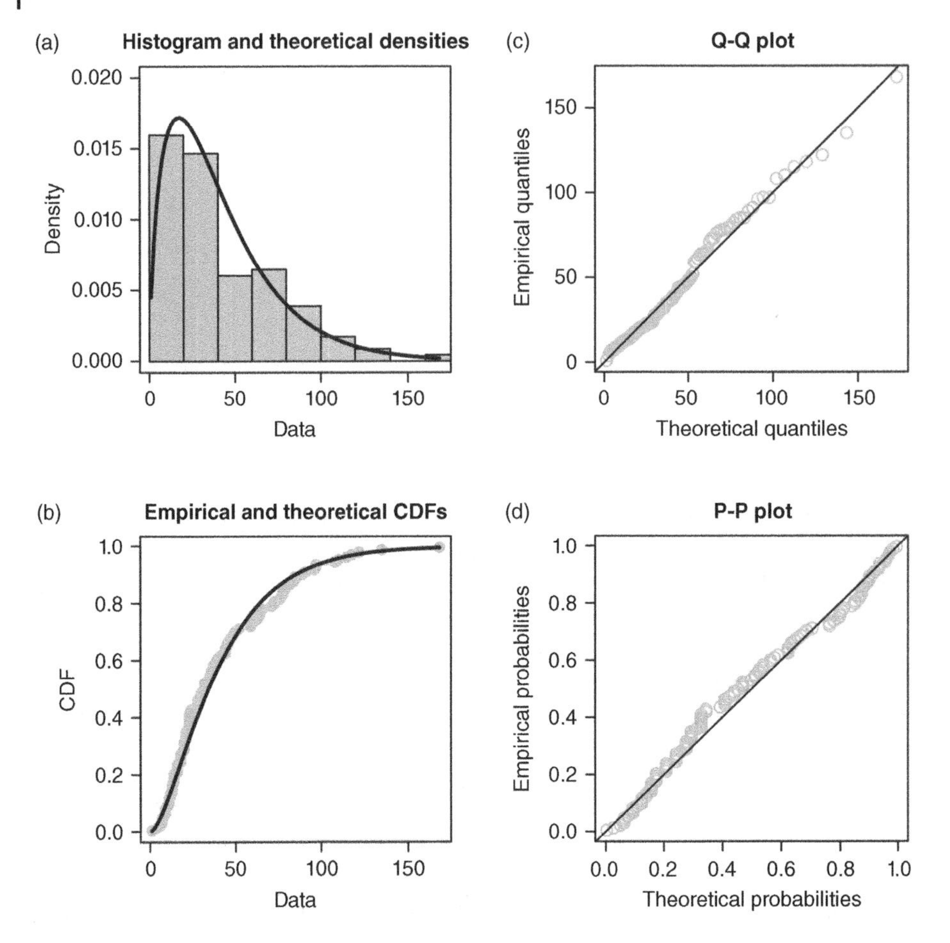

Figure 3.3 (a) Histogram of the ozone concentration from the built-in `airquality` dataset including a probability density curve assuming a gamma distribution, (b) empirical (based on sample data, indicated by black circles) and a theoretical cumulative distribution function (CDF, black line) based on the gamma distribution, (c) quantile-quantile plot (Q-Q plot) displaying the quantiles of our sample data (empirical quantiles) as a function of the theoretical quantiles from a gamma distribution (including a straight reference line passing through the 1st and 3rd quartile), and (d) in an analogous manner a probability-probability plot (P-P plot).

a *P*-value greater than 0.05 ($P > 0.05$) suggests that the sample distribution and theoretical distribution are indistinguishable, and we conclude that the sample data follow the theoretical distribution we tested it against. A GOF test with a *P*-value smaller than 0.05 ($P < 0.05$) indicates that the sample distribution is not consistent with the theoretical distribution.

There are multiple packages available for conducting GOF tests, but the functions in the package `goftest` (Faraway et al. 2021) are perhaps the easiest to apply and understand (see `goftest` helpfile). Below, we show applications of the *Cramér–von Mises* test and the *Anderson–Darling* test using the ozone data (`o3`) we have previously used as an example. First, we specify the distribution function of the assumed distribution (these tests require the cumulative distribution function, which always starts with a 'p' in the R software because it returns the cumulative probabilities, e.g. `pnorm`, `pbinom`). Then, we need to provide estimates of the distribution parameters, which we can derive from the `fitdist` object created previously (which we simply called 'gamma'). We also need to specify that the distribution parameters are estimated from our sample data by setting `estimated = T`, which involves an adjustment for the effect of parameter estimation based on a random data split, so the *P*-value varies with each repetition of the test. In the following example, we use the `set.seed` function to fix the output of the random number generator making sure that we all obtain the same random split and thus the same test results. The so-called 'seed' is a positive integer that provides the starting point of the algorithm generating the random numbers and if we re-run the random number generator with the same seed, we get the same set of random numbers.

Use `set.seed` before any random number generation if you want to reproduce the same numbers every time!

```
## Load required package
> library(goftest)

## Fixing the seed (initialisation) of the random number generator to
   ensure identical random values for all users
> set.seed(seed = 123)

## Cramér-von Mises test
> cvm.test(x = o3, null = "pgamma", shape = gamma$estimate[1],
  rate = gamma$estimate[2], estimated = T)

Cramer-von Mises test of goodness-of-fit
Braun's adjustment using 11 groups
Null hypothesis: Gamma distribution
with parameters shape = 1.69969950295132, rate = 0.0403549556737518
Parameters assumed to have been estimated from data

data:  o3
omega2max = 0.5203, p-value = 0.3122

> ad.test(x = o3, null = "pgamma", shape = gamma$estimate[1],
  rate = gamma$estimate[2], estimated = T)

Anderson-Darling test of goodness-of-fit
Braun's adjustment using 11 groups
Null hypothesis: Gamma distribution
with parameters shape = 1.69969950295132, rate = 0.0403549556737518
Parameters assumed to have been estimated from data

data:  o3
Anmax = 1.5336, p-value = 0.8695
```

Both tests indicate that the null hypothesis cannot be rejected ($P > 0.05$), meaning that our sample distribution does not significantly differ from a gamma distribution. However, we should not get too attached to the outcome of GOF tests for two major reasons: firstly, it is common knowledge that a theoretical distribution cannot perfectly describe real-world data, so we often just want to identify the most closely matching distribution among a set of candidate distributions. Secondly, we urge caution in relation to GOF tests because the power of these tests rises with increasing sample size. At very large sample sizes, even minor departures from the expected distribution can result in statistical significance, potentially detecting differences that are not practically relevant. We will illustrate this with a small example below. We first use a sample of 100 and then a sample of 5000 values randomly drawn from a standard normal distribution in the popular *Shapiro–Wilk* test for normality. The null hypothesis of the *Shapiro–Wilk* test states that a sample comes from a normally distributed population.

So, *P*-values greater than 0.05 indicate that there is no significant departure from a normal distribution.

```
> set.seed(seed = 323) # Ensuring we all have identical random numbers

> shapiro.test(x = rnorm(100)) # 100 random normally distributed values

        Shapiro-Wilk normality test

data:  rnorm(100)
W = 0.98566, p-value = 0.354

> shapiro.test(x = rnorm(5000)) # 5000 random normally distributed values

        Shapiro-Wilk normality test

data:  rnorm(5000)
W = 0.99916, p-value = 0.01577
```

With the smaller sample, the *Shapiro–Wilk* test correctly suggests no statistically significant difference ($P = 0.354$) between the sample distribution and a normal distribution. However, with the large sample size, the normality test detects a statistically significant difference between the sample and the normally distributed population ($P = 0.016$), which would lead us to conclude that our data are not normally distributed, when in fact they are.

3.3 Quantiles

Quantiles are cut-points that divide numeric observations into equally sized groups (subsets) of values when the data are sorted in ascending order. Hence, a quantile marks the boundary between two consecutive subsets. Likewise, they can also be interpreted as dividing a probability distribution into intervals of equal probability. Perhaps, the easiest-to-understand quantile is the median, which divides a series of sorted values into a lower and upper half. The median does not necessarily have to be one of the values in the sample but can take on an intermediate value. The following examples will illuminate this principle: the median of an integer sequence running from 0 to 10 is 5, whereas the median of an integer sequence from 1 to 10 is 5.5. The former is a value that occurs in the original sequence while the latter is not contained in the underlying data.

```
> median(0:10)
[1] 5

> median(1:10)
[1] 5.5
```

In R, quantiles can be derived with the `quantile` function, which allows us to specify one or more cut-points via its `probs` argument (which is short for probabilities). Since the cut-points are given as probabilities, they need to be expressed in the interval between 0 and 1 (the so-called unit interval). This means that the median, which divides our sample into a lower and an upper half, needs to be specified as `probs = 0.5`.

```
> quantile(x = 0:10, probs = 0.5)
50%
  5
```

Quantiles form the cornerstones of the popular boxplot visualisations (see Chapter 6), where a special type of quantiles is used, which are referred to as *quartiles* because they divide the data into four quarters: 25% of the observations lie below the first quartile (Q1), the next 25% lie between the first quartile and the median (second quartile, Q2), another 25% lie between the median and the third quartile (Q3), and the upper 25% of the observations lie above the third quartile (Figure 3.4). The first, second, and third quartiles would be specified as follows:

```
> quantile(x = 0:10, probs = c(0.25, 0.5, 0.75))
25% 50% 75%
2.5 5.0 7.5
```

We could also split our observations into 5 or 10 subsets instead and then obtain *quintiles* or *deciles*, respectively (Figure 3.5). If we use 99 cut-points ranging from 0.01 to 0.99 (1 – 99%),

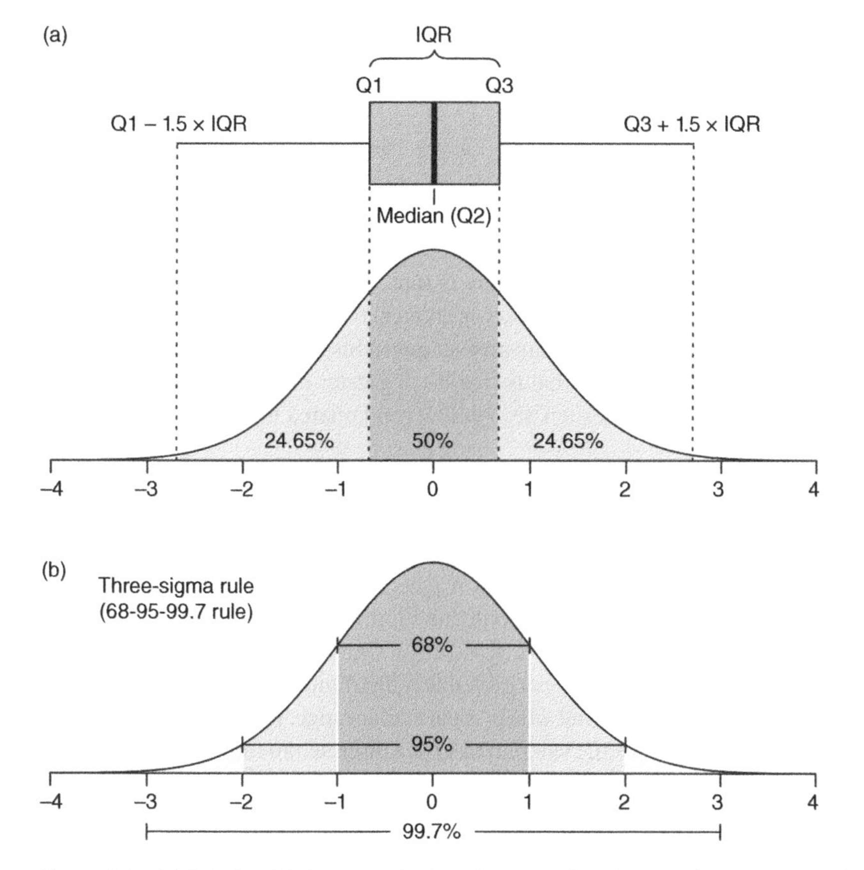

Figure 3.4 (a) Relationship between the boxplot-style visualisation of the quartiles of a random sample drawn from a standard normal distribution with a population mean (μ) of 0 and a population standard deviation (σ) of 1, commonly denoted as $x \sim N$ (0, 1) and (b) the three-sigma rule of thumb stating that *c.* 68% of the observations fall within one standard deviation, *c.* 95% of all observations fall within two standard deviations and *c.* 99.7% of the observations reside within three standard deviations. Q1 = first quartile, Q2 = second quartile, Q3 = third quartile, IQR = interquartile range (Q3 – Q1).

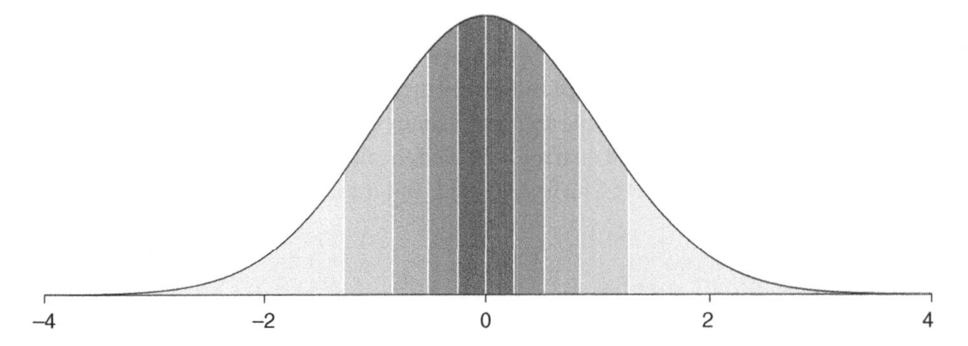

Figure 3.5 Deciles of a standard normal distribution ($\mu = 0, \sigma = 1$) as a specific type of quantile that gives 10 intervals of equal probability. Each of the 10 polygons represents 10% of the total area under the curve.

then the resulting quantiles are called *percentiles* (often used in infant weight-for-age charts for example).

```
## Percentile calculation
> quantile(x = 0:10, probs = seq(0.01, 0.99, by = 0.01))

 1%   2%   3%   4%   5%   6%   7%   8%   9%   10%  11%  12% ...
0.1  0.2  0.3  0.4  0.5  0.6  0.7  0.8  0.9  1.0  1.1  1.2 ...
```

In the context of probability distributions, a quantile simply represents a certain value of a random variable. Now the interesting thing here is that we can derive the probability of randomly drawing a variable whose value is less (or greater) than a certain quantile (simply a user-specified value). In plain English, we can answer questions like: What is the probability of randomly obtaining a measured value that is smaller or greater than a certain threshold?

The following Section 3.4 deals with such probabilities and related topics.

3.4 Probabilities

In the R software, each distribution is associated with a set of four functions revolving around probabilities, whose first letter indicates the specific function, and the remainder is an abbreviation of the distribution name (Table 3.1).

If a random variable approximately follows a probability distribution, then we can calculate the probability of a random variable to fall within a particular range, in case of a continuous random variable, or, to take on a specific value in case of a discrete random variable.

3.4.1 Density Functions (`dnorm, dbinom, ...`)

The `dnorm` function allows us to calculate the probability density of the normal distribution for any given value of our random variable X. In other words, it computes the y-value on the normal curve for a given value of x (multiple x-values can be provided simultaneously). Besides the x-values, the function accepts values for the mean and standard deviation, but these can be omitted in the case of the standard normal distribution (mean of 0 and standard deviation of 1), as this is the default.

Table 3.1 R's built-in family of functions to query probability distributions (here using the example of the normal distribution, but equivalent functions exist for many other distributions).

R function	Description
`dnorm`	**d**ensity function of the normal distribution returns the value of the probability density function (PDF) of the normal distribution for a certain random variable *x*
`pnorm`	**p**robability of a normally distributed random variable *x* to be smaller (or greater) than a specified value (quantile). The probability is derived from the cumulative density function (CDF) of the normal distribution
`qnorm`	**q**uantile function of the normal distribution returns the value of the normally distributed random variable (i.e. the quantile) associated with the user-specified probability (the inverse of the `pnorm` function)
`rnorm`	**r**andom sampling from the normal distribution

```
> dnorm(x = 1) # Equivalent to dnorm(x = 1, mean = 0, sd = 1)
[1] 0.2419707
> dnorm(x = 0)
[1] 0.3989423
```

The probability density associated with a value of 1 in a standard normal distribution is 24% compared to *c.* 40% for the mean at 0.

But wait a minute...we know that the probability of a continuous random variable taking on exactly a certain value is zero. So, what is the conceptual idea behind the probability density then?

Essentially, the probability density provides a *relative likelihood* that the value of the random variable *X* (e.g. biomass, height) would be close to a given *x*. The larger the probability density around a value *x* means that our random variable *X* is more likely to be close to this *x*-value (given the specific mean and standard deviation of the underlying normal distribution).

We also use the `dnorm` function to draw the bell curve associated with a normal distribution:

```
## Plot a bell curve (normal distribution)
> xv <- seq(-4, 4, by = 0.01) # set up an equally spaced vector of x-values
> yv <- dnorm(x = xv) # probability density of the x-values
> plot(xv, yv, type = "l")

## A shorter way to do the same thing...
> curve(expr = dnorm(x, mean = 0, sd = 1), from = -4, to = 4)
```

We use the probability density function of the binomial distribution `dbinom` as an example for discrete variables. The `dbinom` function returns the probability mass of the binomial distribution given a certain random variable *x*, number of trials (`size`), and probability of success on each trial (`prob`). The number of trials refers to so-called Bernoulli trials (or binomial trials), which are experiments with only two possible outcomes: **success** or **failure**. 'Success' in such a trial means that you observe the result you are measuring. For instance, if we are interested in the hunting success rate of predators, then 'success' means that a hunting attempt results in catching prey. Because of the binary nature of the outcome, successes are often indicated by a '1' and failures by a '0'. Based on theoretical considerations or through observations, we can specify success rates as probabilities or the proportion of successes in datasets.

Recall that unlike continuous random variables, we obtain point probabilities with discrete random variables. The `dbinom` function returns such a point probability for a user-specified number of successes, a given number of trials (`size` argument), and the success rate on each trial (`prob` argument). The following example shows how to tackle some interesting probability questions related to binomial variables:

Cheetahs have a hunting success rate of 60%. If such a predator makes 10 hunting attempts, what is the probability of catching prey *exactly* 8 times?

```
## Calculate the probability of exactly 8 successes during 10 trials
   given a probability of success on each trial of 0.6
> dbinom(x = 8, size = 10, prob = 0.6)
[1] 0.1209324
```

The probability of a cheetah catching prey 8 times out of 10 attempts is 12.1%.

3.4.2 Probability Distribution Functions (`pnorm, pbinom, ...`)

The `pnorm` function allows us to determine the probability of an event being smaller (greater) than or equal to a specific value x of a normally distributed random variable. The function uses the underlying CDF to calculate the cumulative probability (see rightmost panels in Figure 3.1). Corresponding functions are available for a wide range of distributions (see the helpfile for `Distributions`). With the help of the `pnorm` function, we can answer practical questions related to continuous variables following a normal distribution. For example, if the average body length of a species of longhorn beetle is 20 mm with a standard deviation of 5 mm, then we can determine the probability of randomly encountering a specimen larger than or equal to 15 mm (Figure 3.6).

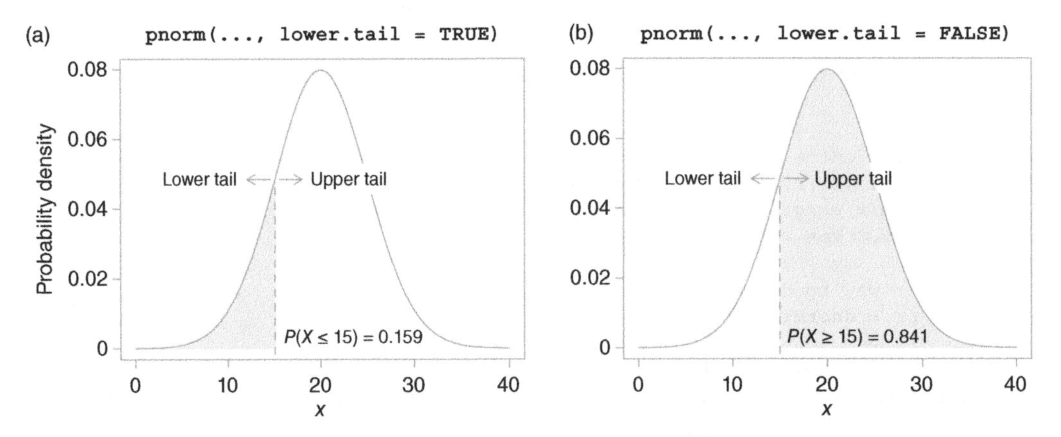

Figure 3.6 Illustration of the lower and upper tail of a distribution using the example of a normal distribution ($\mu = 20, \sigma = 5$, referring to the longhorn beetle example in the text). When computing a probability from the normal distribution using the `pnorm` function, the user needs to be mindful of the `lower.tail` argument. (a) When set to `lower.tail = TRUE`, the `pnorm` function gives the probability of a random sample being smaller than or equal to a certain value (referring to the lower, i.e. the left tail of the distribution indicated by the grey area). (b) When `lower.tail = FALSE`, the probability to the right of a certain value ('greater than or equal to') is calculated (indicated by the grey area).

```
## What is the probability of randomly sampling a beetle ≥ 15 mm given
   a population mean length of 20 mm and a standard deviation of 5 mm?
> pnorm(q = 15, mean = 20, sd = 5, lower.tail = FALSE)
[1] 0.8413447
```

The probability of a random encounter with a beetle ≥15 mm is 84.1%. Because we ask a 'larger than or equal to' question, we need to specify `lower.tail = FALSE`, to make sure that the probabilities related to values ≥15 mm are summed up. The default setting is `lower.tail = TRUE`, which relates to 'smaller than or equal to' questions and results in the cumulative probability of values ≤15 mm (Figure 3.6). So, if we are interested in the probability of randomly coming across a beetle ≤15 mm, we can estimate this as follows:

```
## What is the probability of randomly sampling a beetle ≤ 15 mm?
> pnorm(q = 15, mean = 20, sd = 5, lower.tail = TRUE)
[1] 0.1586553
```

The probability of randomly sampling a 15 mm long or shorter beetle is about 15.9%.

More technically correct, the lower tail refers to the lower values of a distribution, i.e. the area of the probability curve to the left of a certain value. The upper tail contains the higher values of a distribution and therefore indicates the area of the probability curve to the right of a given value (Figure 3.6). In a statistical sense, these values can be seen as cut-points and thus represent quantiles, which explains why the first argument of the `pnorm` function reads `q` (short for quantile).

We can also ask more elaborate questions, e.g. how likely it is to find a specimen between 12 and 17 mm. To this end, we simply subtract the two respective probabilities from each other (the probability of larger value represents the *minuend* and thus comes first, followed by the probability of the smaller value forming the *subtrahend*).

```
> pnorm(q = 17, mean = 20, sd = 5) - pnorm(q = 12, mean = 20, sd = 5)
[1] 0.2194538
```

The probability of encountering a beetle between 12 and 17 mm in length is 21.9%.

We will again use the binomial distribution as an example of a discrete distribution. The `pbinom` function uses the binomial CDF to compute the cumulative probability of observing less (greater) than or equal to a user-specified number of successes (`q`) in n trials (`size`), given a certain success probability (`prob`) on each trial. The term 'trial' refers to a binary observation or event such as a coin flip. This is perhaps best understood with the help of an example:

Cheetahs catch and kill prey in *c.* 60% of their hunting attempts. If a cheetah hunts 10 times, what is the probability that it kills 7 or more times? Please note that we use `dbinom` to calculate probabilities of exact numbers of successes, whereas `pbinom` is used to answer *smaller than or equal to* or *greater than or equal to* questions.

```
## Calculate the probability of more than 7 successes during 10
   trials, with a success probability of 60% on each trial.
> pbinom(q = 7, size = 10, prob = 0.6, lower.tail = FALSE)
[1] 0.1672898
```

The probability of observing more than 7 kills during 10 attempts is 16.7%.

If we ask about the probability of 7 or fewer kills, then we need to set the `lower.tail` argument to `TRUE` (the default).

```
> pbinom(q = 7, size = 10, prob = 0.6, lower.tail = TRUE)
[1] 0.8327102
```

The average cheetah manages 7 or fewer kills per 10 attempts with a probability of 83.3%.

It is also possible to obtain the probability for a range of kills. We can ask, for example, what is the probability of a cheetah catching between 3 and 5 prey during 10 trials?

```
## Calculate the probability of observing between 3 and 5 successes
   during 10 trials, given a success rate of 60% on each trial.
> pbinom(5, size = 10, prob = 0.6) - pbinom(3, size = 10, prob = 0.6)
  [1] 0.3121349
```

The probability of a cheetah catching between 3 and 5 prey is 31.2% (given the 0.6 success rate at each trial).

3.4.3 Quantile Functions (`qnorm, qbinom, ...`)

The quantile functions like `qnorm`, `qgamma`, `qpois`, etc. provide the inverse of the probability distribution functions (inverse of the CDF), e.g. the `qnorm` function represents the opposite of the `pnorm` function. In other words, we use the quantile functions when we are interested in the value of x that is associated with a certain probability (x represents a quantile). Sticking with the previous longhorn beetle example (see Section 3.4.2), we can determine the quantile below which 30% of the beetle lengths fall.

```
> qnorm(p = 0.3, mean = 20, sd = 5)
[1] 17.378
```

The outcome tells us that 30% of the individuals of this beetle species show lengths below 17.4 mm (note that we do not need to specify '`lower.tail = TRUE`' since this is the default). When specifying '`lower.tail = FALSE`', we obtain the quantile above which 30% of the beetle lengths lie.

```
> qnorm(p = 0.3, mean = 20, sd = 5, lower.tail = F)
[1] 22.622
```

The body length of 30% of the beetle population is equal to or exceeds 22.6 mm.

With the help of the quantile functions, we can also calculate the range of values within which 95% of the observations lie. Figuratively speaking, we need to chop off 2.5% at the lower and upper tail of the distribution to obtain the central range comprising 95% of the values. For our longhorn beetle species, this results in the following R code:

```
## Calculate the interval within which 95% of the observations fall
> qnorm(p = c(0.025, 0.975), mean = 20, sd = 5)
[1] 10.20018 29.79982
```

Our calculations show that the central 95% range of longhorn beetle lengths spans values from *c.* 10 to 30 mm.

Continuing with our discrete cheetah example of a binomial distribution, we can use `qbinom` to compute the quantile for a given probability (or a vector of probabilities). The output of the `qbinom` function is the number of successes associated with the cumulative probability. Essentially, `qbinom` performs the inverse of the `pbinom` function.

We can ask questions like: How many cheetah kills will we see 25% of the time (i.e. if we repeatedly observe sets of 100 hunting attempts with a success rate of 0.6 on each trial)?

```
> qbinom(p = 0.25, size = 100, prob = 0.6)
[1] 57
```

We will see 57 or fewer cheetah kills 25% of the time (given the specified number of trials and associated success rate per trial).

We can also use it to quantify the central 95% range of successes (here cheetah kills).

```
> qbinom(p = c(0.025, 0.975), size = 100, prob = 0.6)
[1] 50 69
```

So, 95% of the time, we will witness between 50 and 69 cheetah kills during 100 trials (given a 0.6 success rate per trial).

3.4.4 Random Sampling Functions (`rnorm`, `rbinom`, ...)

The `rnorm` function generates a set of random observations drawn from a normal distribution, which we can use to simulate data for learning and model training purposes as well as for testing statistical models. The first function argument `n` allows us to specify the number of values we would like to draw randomly from a normal distribution. As before, leaving out the arguments `mean` and `sd` assumes a standard normal distribution by default.

```
> set.seed(seed = 123) # Ensures we all get the same set of random values
> rn <- rnorm(n = 1000)
> head(rn) # view the first few values
[1] -0.56047565 -0.23017749  1.55870831  0.07050839  0.12928774  1.71506499
> hist(rn)
```

So, if we want to simulate body lengths of the longhorn beetle species we used in the previous examples (Sections 3.4.2 and 3.4.3), we need to specify the mean and standard deviation.

```
> set.seed(seed = 123)
> nbeetle <- rnorm(n = 1000, mean = 20, sd = 5)
> hist(nbeetle)
```

Many processes or observations in nature can be modelled as independent Bernoulli trials (presence/absence, dead/alive, etc.) using the binomial distribution. We can draw random numbers from this discrete distribution using the `rbinom` function, whose first argument is also called `n` and allows us to specify the number of desired random observations. The confusing thing here is that the number of random observations refers to how many *sets* of Bernoulli trials (of a given size) should be generated or in other words, how many times R should simulate an experiment. The size of these Bernoulli sets, however, is specified via the `size` argument and can be as little as one, e.g. a single flip of a coin. Finally, we encounter the `prob` argument again, which specifies the probability of success on each trial. Considering our cheetah example, we could simulate our set of 10 trials a hundred times like this:

```
> set.seed(seed = 123)
> rb <- rbinom(n = 100, size = 10, prob = 0.6)
> rb
  [1]  7  5  6  4  4  9  6  4  6  6  3  6  5  6  8  4  7  9  7  3  4  5  5  2
 [25]  5  5  6  6  7  8  3  4  5  5  9  6  5  7  7  7  8  6  6  7  8  8  7  6
 [49]  7  4  9  6  5  8  6  7  8  5  4  7  5  8  6  7  5  6  5  5  5  6  5  6
 [73]  5 10  6  7  7  6  7  8  7  5  6  5  8  6  3  4  4  7  8  5  7  5  7  7
 [97]  5  8  6  6
```

For ease of interpretation, we go through the first few numbers: the first value means that 7 hunting successes occurred in the first simulated set of 10 attempts. In the second set of 10 attempts, our cheetah caught 5 prey while there were 6 successes in the third simulated set of

10 attempts, and so on. We can easily summarise this information in a one-way contingency table (frequency counts for a single variable).

```
> table(rb)
rb
 2  3  4  5  6  7  8  9 10
 1  4 10 23 24 21 12  4  1
```

This overview quickly tells us that 2 hunting successes (out of 10 attempts) occurred in only one of the 100 simulations, while 3 hunting successes were seen in 4 of the 100 simulations, and so on. In accordance with the specified success rate of 0.6 per trial, 6 hunting successes (6/10 = 0.6) occurred most frequently among the 100 random simulations of 10 attempts each.

3.5 The Normal Distribution

The normal (Gaussian) distribution is a bell-shaped distribution that is symmetric around the mean (μ, location parameter) and its width is determined by the standard deviation (σ, spread parameter). Because of the continuous nature of the underlying random variable and the two defining parameters, the normal distribution is more generally called a two-parameter continuous probability distribution. The symmetrical nature means that the mean, median, and mode of a normal distribution are identical. Importantly, the area under normal curve is equal to 1 (100%), so the area between two points gives the probability of values falling into this range. The normal distribution is the most famous and widely used distribution in statistics. Various biological and environmental variables roughly follow a normal distribution and therefore it plays an important role as the assumed default distribution of the response variable in many statistical models (Figure 3.1).

There is a nice rule of thumb, called the *three-sigma rule*, which states that 68% of the possible values lie within $\pm$ 1 standard deviation of the mean, 95% of the values lie within ± 2 standard deviations and 99.7% of the values are within ± 3 standard deviations (Figure 3.4). We can easily deduce this rule using the probability density function for the normal distribution `pnorm`. Leaving the mean and standard deviation unspecified, this function assumes a standard normal distribution, i.e. a normal distribution with a mean of zero and a standard deviation of one ($\mu = 0$ and $\sigma = 1$).

```
> pnorm(q = 1) - pnorm(q = -1)
[1] 0.6826895
> pnorm(q = 2) - pnorm(q = -2)
[1] 0.9544997
> pnorm(q = 3) - pnorm(q = -3)
[1] 0.9973002
```

3.6 Central Limit Theorem

The central limit theorem (CLT) states that the distribution of *sample means* (not the raw data!), regardless of the distribution of the underlying population, approximates a normal distribution (Figure 3.7). The CLT further specifies that the average of the sample means

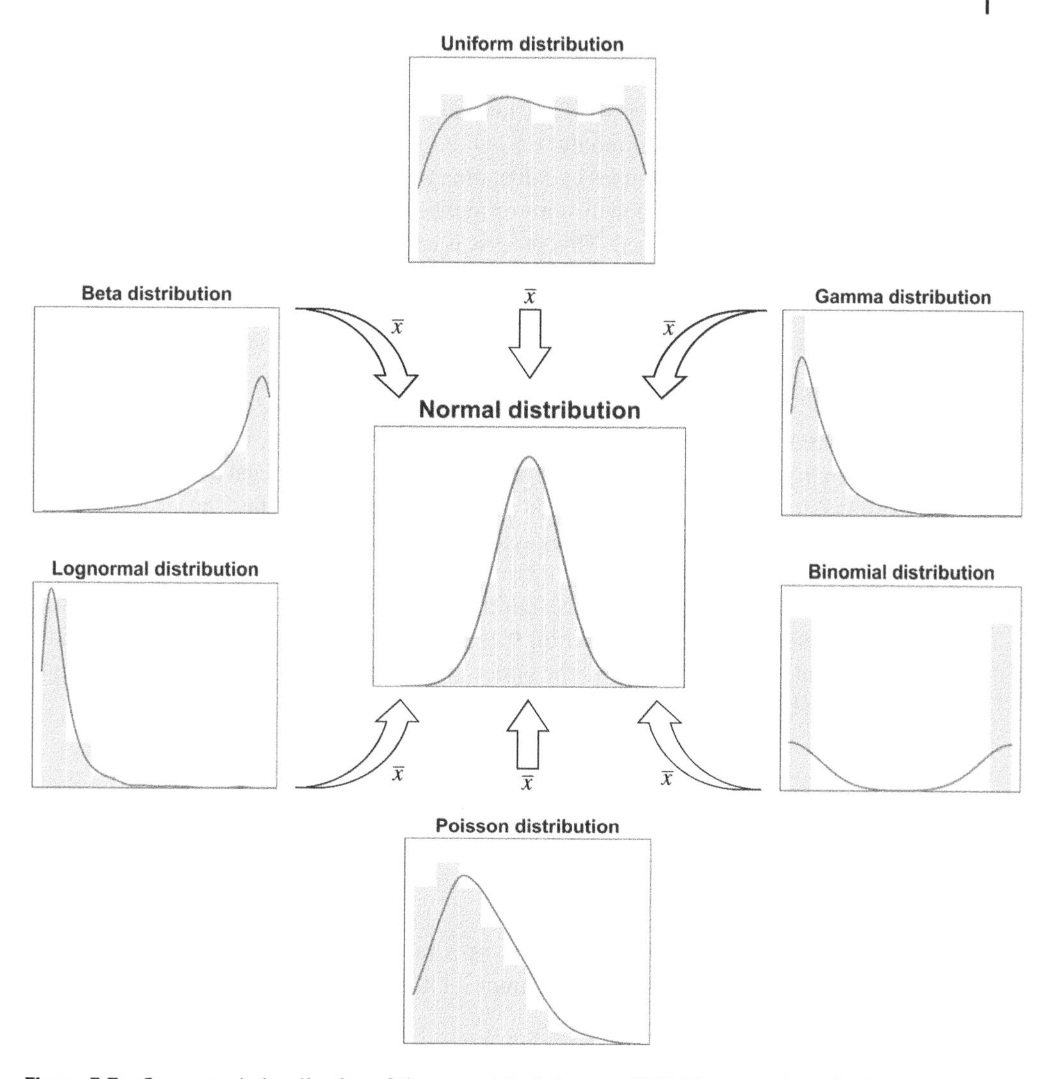

Figure 3.7 Conceptual visualisation of the *central limit theorem* (CLT). The central graph shows a histogram of a set of sample means ($\bar{x}$) originating from any distribution. Given a reasonably large sample size ($n \geq 30$), the distribution of *sample means* of random variables approximates a normal distribution, regardless of the underlying distribution. Please note that the validity of this theorem is not limited to the exemplary distributions shown here.

(and standard deviations) will equal the population mean (and standard deviation). The CLT assumes independent, sufficiently large samples ($n \geq 30$) of a random variable.

So, if we repeatedly take samples from, say, a uniform distribution, calculate the mean of each sample and use any of the graphical tools for distribution checking (e.g. Q-Q plot), then we should see a close alignment between the distribution of the sample means and a normal distribution. The following R code provides such an example, where we take 1000 samples from a uniform distribution with an individual sample size of 50. Then we apply the previously introduced `fitdist` function and plot the results (figure not shown).

```
> uniform <- replicate(n = 1000, expr = mean(runif(n = 50)))
> check <- fitdist(data = uniform, distr = dnorm)
> plot(check)
```

Why is it important to know about the CLT?

The CLT is a cornerstone of modern statistics because it permits us to safely assume that the sampling distribution of the mean of any random variable will typically follow a normal distribution. As scientists, we commonly work with samples rather than the entire population, and we often summarise samples by calculating the mean. On the basis of these sample means, we usually aim to draw conclusions or make statements about the population (i.e. we wish to generalise our findings). This process is called *statistical inference* and involves using statistical techniques to make predictions or estimates about a population based on a sample. This means that the CLT can be used to construct confidence intervals around a sample mean (usually a 95% confidence interval) or to test hypotheses about the mean of a population.

A 95% confidence interval is a range of values that is likely to include the true population parameter (e.g. the mean) with a probability of 95%. In other words, if a sufficiently large sample is taken from a population and a confidence interval is constructed using that sample, the interval will include the true population parameter 95% of the time. The CLT allows us to take advantage of statistical techniques assuming a normal distribution such as the fact that approximately 95% of the values in a standard normal distribution lie roughly within ± 2 (Figure 3.4). This is the reason why we multiply the standard error of the mean with the quantiles delineating the interval containing 95% of the values. To obtain such a 95% interval, we need to exclude the lower 2.5% and the upper 2.5% of the possible values (i.e. we confine the sample space between 2.5% and 97.5%).

```
## Compute the quantiles corresponding to 2.5% and 97.5%
   probability. These quantiles encompass the interval within which 95%
   of the values lie
> qnorm(p = 0.025)
[1] -1.959964
> qnorm(p = 0.975)
[1] 1.959964
```

This approach yields the well-known 1.96 as a multiplication factor for the standard error to obtain the lower and upper bounds of the 95% confidence interval around the mean (see Sections 12.1 and 12.3).

3.7 Test Statistics

Test statistics are often simply scaled versions of the effect size. The effect size can be the difference in means, the steepness of a slope, the strength of a correlation between two variables, or the risk of an event that may occur. By 'scaled', we mean that the raw effect size is divided (standardised) by a measure of uncertainty. The underlying idea is that a meaningful test statistic should take the uncertainty around an effect into account. Consequently, strong effects with little uncertainty yield large test statistics but in case of no effect or if the effect gets lost in the 'noise', we obtain small test statistics. For example, in a *t*-test, which is used to compare a continuous response variable among two groups, the test statistic is calculated as the difference in means between the two groups (effect magnitude) divided by their pooled standard deviation (dispersion or spread). Another example is linear regression, where the value of the slope represents the effect size (steepness of the slope) and dividing it by its standard error (a measure of uncertainty) gives the test statistic.

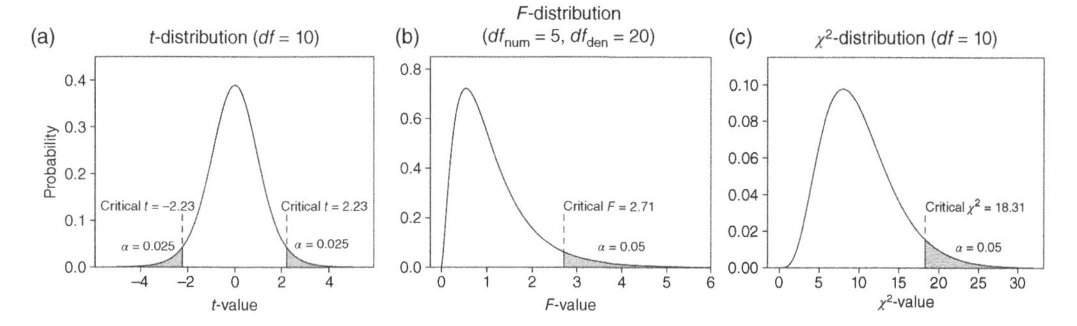

Figure 3.8 Common test statistics and their distributions. (a) t-distribution with 10 degrees of freedom (*df*), the t-statistic is linked to Student's t-test, which also operates in various regression models, (b) F-distribution of the F-statistic associated with analysis of variance and related statistical models, (c) χ^2-distribution related to the χ^2-statistic of goodness-of-fit tests and the likelihood ratio test statistic.

In both examples, the resulting test statistic follows a t-distribution. We can now query the t-distribution about the probability of obtaining a test statistic of this magnitude or greater just by chance (here magnitude refers to the absolute value of the test statistic regardless of sign). If the resulting probability is smaller than a certain probability level (usually 0.05), then we conclude that there is a statistically significant effect (e.g. a statistically significant difference between means in a Student's t-test or a slope that is significantly different from zero in a linear regression context). In this way, we can also determine critical values of a test statistic, i.e. the threshold that defines the boundary of the rejection region (the region where the probability drops below 0.05) (Figure 3.8). Values of the test statistic that lie in the rejection region have a very low probability to occur by chance and thus prompt us to assume that a true effect exists (we reject the null hypothesis of no effect). The t-distribution is symmetric (very similar to the normal distribution) and by default often used in support of two-tailed tests, where two critical values exist, one in the lower tail and one in the upper tail, allowing to test 'greater than or equal to' or 'smaller than or equal to' hypotheses. Other frequently encountered distributions are the F-distribution associated with the F-statistic used in analysis of variance and related models (Chapters 10 and 11) as well as the well-known chi-square (χ^2) distribution linked to the χ^2-statistic in chi-square tests of goodness-of-fit or chi-square test of independence (see Chapter 7). Perhaps most importantly, the test statistic of the classical and broadly applied likelihood ratio tests used for comparing nested models, approximately follows a chi-square distribution. Both the F- and chi-square distributions are non-symmetric and mostly used in one-tailed tests (Figure 3.8).

3.7.1 Null and Alternative Hypotheses

Classical statistical testing begins with the formulation of the null hypothesis (H_0) and the alternative hypothesis (H_A), two mutually exclusive statements about a population. However, because it is commonly not possible to work with the entire population, the associated hypothesis tests rely on sample data to determine whether to reject the null hypothesis (and thus accept the alternative hypothesis). The null hypothesis (H_0) assumes no effect, no difference between groups or no relationship between variables. When we talk about effects, we generally refer to experimental treatments or environmental effects. Groups can, for instance, be different species, functional groups, habitats and so on that we want to compare. The alternative hypothesis (H_A) assumes an effect, a difference of some sort or a relationship between variables. We can distinguish two basic H_A scenarios:

- A *two-sided test* (two-tailed test) assumes an effect or a difference without stating a direction (positive or negative effect, the variable of interest in group A is higher or lower than in group B). In other words, with this type of test, we leave it open if a treatment has a positive or negative effect (see Section 5.2.1.3 and Figure 5.5).
- A *one-sided test* (one-tailed test) assumes that the effect or the difference between groups has a direction (either greater or smaller; see Section 5.2.1.3 and Figure 5.5).

Most statistical analyses are performed as two-sided tests by default but with tests that allow us to choose between one- and two-sided options, we should do so before seeing the data. Opting for a one-sided test requires a valid justification, i.e. we should be certain that an outcome can only go in one direction based on physical or physiological reasons. For example, exposing plants to low and high light under otherwise similar non-limiting conditions would almost certainly lead to greater growth in the plants subjected to high-light conditions. In this case, it would be justified to run a one-sided test assuming significantly larger plants under high-light conditions.

3.7.2 The Alpha Threshold and Significance Levels

The basic idea of hypothesis tests is to derive a test statistic. These test statistics follow certain probability distributions, which means we can query the underlying distribution of how likely it is to obtain a test statistic as extreme or more extreme than the value obtained with our statistical test or model, just by chance (the term 'extreme' is used here to cover both scenarios 'as small or smaller than' or 'as large or larger than'). The probability associated with a test statistic is called the *P*-value. Now the question of the threshold for this *P*-value lingers, i.e. how small this *P*-value should be, for us to reject the null hypothesis and conclude that there is a significant effect. This threshold value is referred to as the significance level or alpha level (α). The significance level is commonly (arbitrarily) set to 5% ($\alpha = 0.05$). So, we compare the *P*-value of our test statistic to our significance level and if *P* is smaller than 0.05, we reject the null hypothesis and accept the alternative hypothesis. At the same time, this means that we allow ourselves a 5% error, or, in other words, there is a 5% probability of making the wrong decision when the null hypothesis is actually true.

3.7.3 Type I and Type II Errors

When we set the level of significance for a hypothesis test to $\alpha = 0.05$, then we automatically accept a 5% error probability, i.e. purely by chance, 5% of the time we incorrectly conclude that there is a statistically significant effect such as a difference in two population means (Figure 3.9). This is called a ***type I error*** and implies that with an α of 0.05, on average 1 out of 20 statistical tests turns out statistically significant purely by chance, even though there is no true effect or difference between groups in reality (Figures 3.9 and 3.10, Table 3.2). If we want to lower the risk of making a type I error, we must lower our significance level α. However, this is a two-edged sword because the lower we set the value for α, the less likely we are to detect a true effect/difference if it really exists.

A ***type II error*** occurs when the null hypothesis is false, but we fail to reject it. In other words, with a type II error, we incorrectly conclude that there is no effect/difference, even though it actually exists (Figures 3.9 and 3.10, Table 3.2).

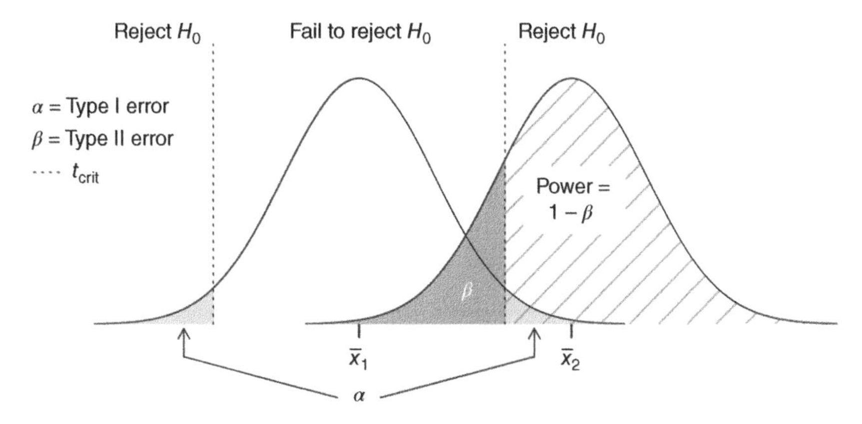

Figure 3.9 Type I error (**α**, light-grey areas), type II error (**β**, dark-grey area), and statistical power (1 − **β**, hatched area) of a two-sample, two-tailed *t*-test. The curves represent the distributions of samples coming from two groups. The dotted lines indicate the critical *t*-values beyond which the test rejects the null hypothesis (no difference in means). Imagine there is a true difference in population means between the two groups, but purely by chance our random sample for group 2 falls largely into the dark-grey range of possible values. Due to this sampling bias, we would underestimate the mean of population 2 and thus the effect size, resulting in a smaller *t*-statistic than the critical value. In this scenario, we would commit a type II error (*false negative*) because we were unable to reject the false null hypothesis yielding a non-significant *t*-test result. Conversely, if no true difference in group means existed, but purely by chance our sample of group 2 would largely consist of values to the right of the dark-grey area, then we would overestimate the effect size and commit a type I error, i.e. falsely rejecting the null hypothesis and concluding that there is a statistically significant difference between group means. $\bar{x}_1$ = population mean of group 1, $\bar{x}_2$ = population mean of group 2.

Figure 3.10 Cartoon illustrating type I (false positive) and type II (false negative) errors.

Both types of error are 'bad' of course, but depending on the case, there may be reasons why one should be trying to minimise a type I or type II error. Consider a scenario where the efficacy of a treatment against cancer is tested but it is known that the treatment has severe side effects. In this case, you would want to be absolutely sure that your treatment provides a benefit to your patients, knowing that they will have to cope with the adverse side effects.

Table 3.2 Overview over sample-based decision scenarios with reference to the population reality.

Sample-based decision	True situation in the population	
	H_0 is true	H_0 is false
Reject H_0	***Type I error*** – rejecting H_0 when it is true (probability = α)	Correct decision (probability = $1 - \beta$)
Fail to reject H_0	Correct decision (probability = $1 - \alpha$)	***Type II error*** – fail to reject H_0 when it is false (probability = β)

The level of significance (α) equals the type I error probability (rejecting the null hypothesis when it is true)

So, to minimise a type I error (concluding that the treatment works while in reality it does not, or, in other words, by chance getting a sample that lies to beyond the α-threshold), you would be advised to set the α-threshold as low as possible, e.g. to 1% ($\alpha = 0.01$) rather than the usual 5% ($\alpha = 0.05$). On the other hand, if you were to test whether an optional prenatal diagnostic may have a fatal side effect on the embryo, you would want to avoid a type II error (concluding that there is no risk of a fatal side effect while in reality there is one). Under these circumstances, it is advisable to raise the α-threshold to maximise the so-called 'power' of the test to detect a difference.

References

Delignette-Muller, M.L. and Dutang, C. (2015). fitdistrplus: An R package for fitting distributions. *Journal of Statistical Software* 64 (4): 1–34. https://doi.org/10.18637/jss.v064.i04.

Faraway, J., Marsaglia, G., Marsaglia, J., and Baddeley, A. (2021). goftest: Classical goodness-of-fit tests for univariate distributions. *R Package Version 1.2-3*. https://CRAN.R-project.org/package=goftest.

4

Replication and Randomisation

In Chapters 2 and 3, we have looked at variation, where it originates from, and how we can characterise it, both visually and numerically. We have also seen how to distinguish between different types of variation – *random* and *systematic*. Understanding the reason for systematic variation provides us with explanatory power, and we achieve this by trying to maximise the signal-to-noise ratio. Two fundamental tools that help us with the causal assignment of variance are *replication* and *randomisation*. Even a slight misunderstanding of these concepts can cause ripple effects far into your statistical analyses. Both are intricately related, and both also relate to the concept of statistical independence. To fully understand Boxes 4.1 and 4.2, you will have to read Chapter 5, but it should be possible to follow the argumentation without doing so.

4.1 Replication

Whilst the meaning of the word 'replication' may be clear to us in our day-to-day use of the English language, we need to understand what it means in a statistical sense. Consider this example:

You have three plants in three identical pots, and with homogenous soil (Figure 4.1a). To one pot, you add nitrogen (N), to another one you add N and water, and the third one remains as a control (see Chapter 2). After a certain time, you measure the height of the plants. A statistically uninformed experimenter might argue that the plant in the middle of Figure 4.1a grew best because it received additional nitrogen and water. The problem with this statement is that there is absolutely no way to tell apart the *natural* (inherent, preexisting) unsystematic variation between the three plants, and that imposed by the treatments (the systematic variation caused by nitrogen and water). This is because the (very slight, unknown) differences between the three experimental units (in terms of the soil, plant seeds, pots, environmental conditions, etc.) *coincide* with the three treatment conditions. For example, if the first pot, which has received nitrogen, at the same time happens to be exposed to slightly more light, then the two effects are irreversibly confounded. We will never be able to causally explain why this particular plant grew more – it could be due to the added nitrogen – or due to the slightly higher light or any other inherent properties of that pot or plant. One might argue that, if extreme care is taken, we can ensure that the conditions between the three pots are exactly the same. This is easily countered by growing a number of plants under

R-ticulate: A Beginner's Guide to Data Analysis for Natural Scientists, First Edition.
Martin Bader and Sebastian Leuzinger.

Companion website: www.wiley.com/go/Bader

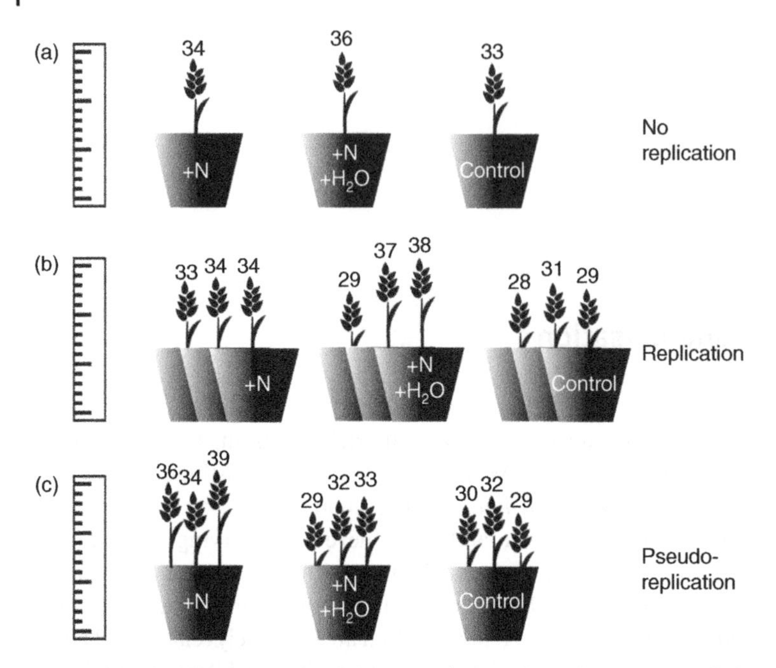

Figure 4.1 An illustration of a simple experiment where the response of plant height to nitrogen (N) and water (H_2O) addition is tested. (a) A design that lacks replication ($n = 1$), one with minimal replication (b), and a design showing pseudoreplication (c). Numbers above plants indicate their height in cm.

Replication means we are assigning more than one experimental unit to a particular treatment.

Pseudo-replication occurs if we erroneously consider dependent samples as independent.

The number of replicates is commonly indicated by an italicised lower-case letter *n*.

the exact same conditions – even then, we will observe that the plant heights differ. We have to think of effects as subtle as the movement of air in an air-conditioned room or an invisible pathogen that lives inside a plant. This is equally true for non-living subjects, namely the pots that we are using in the above example. Despite the fact that they look the same, there will be very slight variation around their mass, volume, perhaps slight contamination, etc. Therefore, the experiment shown in Figure 4.1a is *not* replicated, and thus unfit to provide any meaningful answer to the question of what N and water addition cause in those plants.

In Figure 4.1b, we use three pots per treatment. As long as we can ensure their independence (we will see what exactly this means), we now have a replicated experiment, where three plants each receive the same treatment, namely nitrogen, both nitrogen and water, or nothing (control). Our *unit of replication* is the plant, and because we have three of those units per treatment, we say that our '*n*' is three. Just by looking at Figure 4.1b, you can get a visual impression of the variation in height between the plants: the three control plants all seem to be around 30 cm tall, seemingly less than the plants that received nitrogen, and definitely less tall than the plants that received both water and nitrogen. (Whether these apparent differences can be considered 'significant', and how to test for this, is another question, which we will answer in the following chapters.) For now, it is critical to see that if we assign *multiple* units of replication (at least three) to the treatments as well as the control, this provides us with an opportunity to distinguish between the *signal* (systematic variation, the treatment effect), and the *noise* (unsystematic variation), i.e. the variation within a particular treatment or control.

An economically conscious researcher might suggest planting the three replicates per treatment into the *same* pots (Figure 4.1c), arguing that this will save resources and produce a similar result, as we can still distinguish between signal and noise. However, this is rather wrong and leads to a dangerous phenomenon called *pseudoreplication* (we intend to replicate, but we do not do it properly). The problem is that the three plants within one pot are *not independent*. If, for instance, the middle pot leaks a small amount of water or contracts a plant pathogen, *all* three plants within that pot are affected. This makes the unit of replication no longer the plant (as in Figure 4.1b), but the pot, such that we are back to $n = 1$, as we only have one independent unit per treatment. What are the consequences if pseudoreplication occurs? You can see from Figure 4.1c and Box 4.1 that we risk to interpret the unknown 'pot effect' as a treatment effect. Or, equally bad, a pot effect can blur a treatment effect, as all plants in that pot are affected, such that again, just like in Figure 4.1a, the treatment effect is confounded by the pot effect.

Box 4.1 Simulating pseudoreplication in R

To truly understand the implications of pseudoreplication, let us simulate the phenomenon in R. Assume we are again dealing with a fertiliser experiment where we have four control pots and four fertilised pots, each with five individuals of the same grass species. Because we simulate the data, we can decide whether we want a difference in biomass between the two groups or not, i.e. if we want to see a fertiliser effect. Let us have no effect, the mean for all pots equals 5. The units are arbitrary and reflect the biomass of the plants. To mimic the dependence within pots, we add the same random value to all measurements that were taken within one pot. This reflects the 'pot effect'. The variable `poteffect` contains these eight values, which are *unique* to each pot, but affect all five individuals of a pot. The pot effects could, for example be due to the small differences between the pots, the soil, the environmental conditions, or the biotic interactions within pots. The key is that they are small but neither predictable nor measurable:

```
poteffect <- rnorm(8, mean = 1)
## The first four pots are not fertilised (control)
## Use the set.seed command to obtain the same sets of random numbers
set.seed(11)
pot1 <- rnorm(5, mean = 5) + poteffect[1]
pot2 <- rnorm(5, mean = 5) + poteffect[2]
pot3 <- rnorm(5, mean = 5) + poteffect[3]
pot4 <- rnorm(5, mean = 5) + poteffect[4]

## The second four pots are fertilised (treatment)
pot5 <- rnorm(5, mean = 5) + poteffect[5]
pot6 <- rnorm(5, mean = 5) + poteffect[6]
pot7 <- rnorm(5, mean = 5) + poteffect[7]
pot8 <- rnorm(5, mean = 5) + poteffect[8]
```

Let us visualise the biomass of the grass plants inside the eight pots. Every bar represents the biomass measurement of a single plant, the grouping shows the eight pots. The black

(Continued)

Box 4.1 (Continued)

arrows represent the random pot effects, which either pull the mean of all five plants up or down:

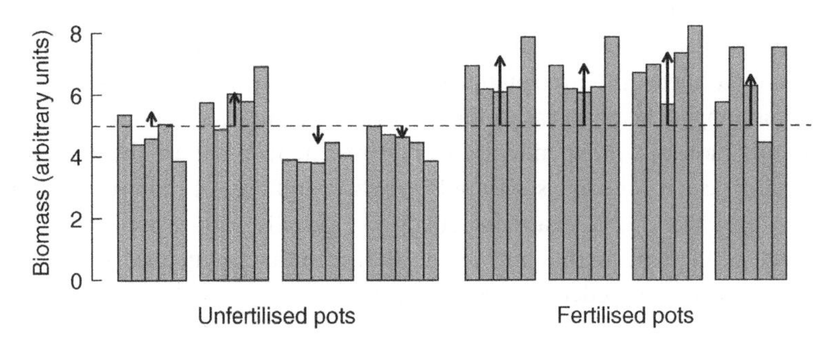

Remember (and you can easily see this from the code above) that we do not simulate a fertiliser effect (the means per pot before the addition of the pot effect are all 5, see the dashed line). Does it look like there is a fertiliser effect? Yes! It looks like there is a small effect. This is purely because the random pot effects on the control side seem small or negative, while those of the four treatment plots happen to be positive. Pseudoreplication arises if we consider every grass plant as a replicate. This is obviously wrong, as it inflates the random pot effects. The experimental design is not flawed as such, but for a correct analysis, we would need to consider the *pot* our unit of replication, and not the grass plant.

To demonstrate this in numbers, we will analyse this dataset once correctly, and once incorrectly (using pseudoreplication). You will have to read Chapter 5 to understand how to conduct and interpret *t*-tests, which are used to compare the means of two groups.

```
## Pseudoreplicated t-test:
t.test(c(pot1, pot2, pot3, pot4), c(pot5, pot6, pot7, pot8))
## The t-test finds a significant difference! (P-value < 0.05)

## Correct t-test, using the means per pot:
t.test(c(mean(pot1), mean(pot2), mean(pot3), mean(pot4)),
   c(mean(pot5), mean(pot6), mean(pot7), mean(pot8)))
## The t-test finds no significant differences between treatments!
```

You could argue that this is a peculiarity of our specific random number generation (using `set.seed(11)`. However, if you repeat the erroneous analysis many times on newly generated datasets, you will see that you systematically commit a higher than expected number of type I errors. Analogously, if the groups were in fact different, you would commit an unusually high number of type II errors. In fact, type I and type II error inflation is one of the primary dangers of pseudoreplication, so watch out for it!

4.2 Statistical Independence

One can argue that while we have three separate pots per treatment in Figure 4.1b, those three pots are still not independent, as they are placed on the same table, in the same room, within the same building, etc. Imagine, for example that some pathogen gets spread through air circulation within that building. In that case, some or all nine pots might be affected, and we could not argue that the pots are independent units of replication. Of course, this argumentation can be continued *ad absurdum*, and will only stop once we realise that there is only one planet to conduct experiments on. Naturally, it would be nonsensical to argue that way, but it shows that the level at which we consider our units of replication independent is somewhat fuzzy. In Figure 4.1, having three plants within a single pot is a clear case of pseudoreplication, as the 'pot effect' applies equally to all three plants (see Box 4.1). Is it therefore wrong to have three plants per pot in our example? The answer to this question is no, but because the level of replication is the pot, we will still have to have three pots per treatment. We can then average the plant height (or biomass) per pot (which will provide us with a more stable mean) and use these mean values rather than all individual measurements (see Box 4.1). We are *averaging pseudoreplication away* so to speak.

Multiple measurements per unit of replication need to be averaged and not mistaken as replicates.

Considerations of statistical independence are important but rarely absolute. Both experimental and observational studies are necessarily confined spatially and temporally and often need to be confirmed via additional, independent data. Even entire studies from the same institution, laboratory, or researcher can be considered not entirely independent. This does not render them invalid, but additional confidence in the results is gained by confirmatory studies from different countries, laboratories, people, and so forth. Generally, we aim to achieve the most robust design given the logistical, financial, and spatiotemporal constraints. Higher replication (or a higher 'n') will always result in a higher signal-to-noise ratio, and therefore higher chances to find treatment effects, should they be present. Identifying the level at which we can consider our experimental units to be independent is the key task. Often, we can improve statistical independence without added cost if we are aware of the problem, for example not only while choosing our subjects (e.g. patients) or experimental units (pots) but also while planning observational studies. Subjects in close spatial or temporal vicinity cannot be considered independent as they could be affected by a particular condition common to an area or time period. For example, if the quality of mass fabricated products are tested, we make sure not to take them from the same batch, produced on the same day, etc.

Replication and statistical independence are pivotal concepts in statistical hypothesis testing, they are tightly coupled – replication ensures statistical independence, and pseudoreplication ultimately is caused by a lack thereof. Randomisation is a further tool to ensure statistical independence, we will discuss this next.

4.3 Randomisation

Let us elaborate on the above example, where we look at the effect of nitrogen and water on plant height. Imagine we are conducting the experiment using proper (albeit minimal) replication as shown in Figure 4.1b. We place our nine pots into the experimental chamber as shown in Figure 4.2a, so ordered by treatment. The schematic shows that the room has a window on one side, but it could also be an air conditioning outlet, a shelf, or simply a wall. The point here is that the experimenter decided to arrange the plants in rows of the same treatments. Intuitively, this might make sense, particularly if the experimental chamber looks homogenous at first, and perhaps such an arrangement also facilitates water and nitrogen addition.

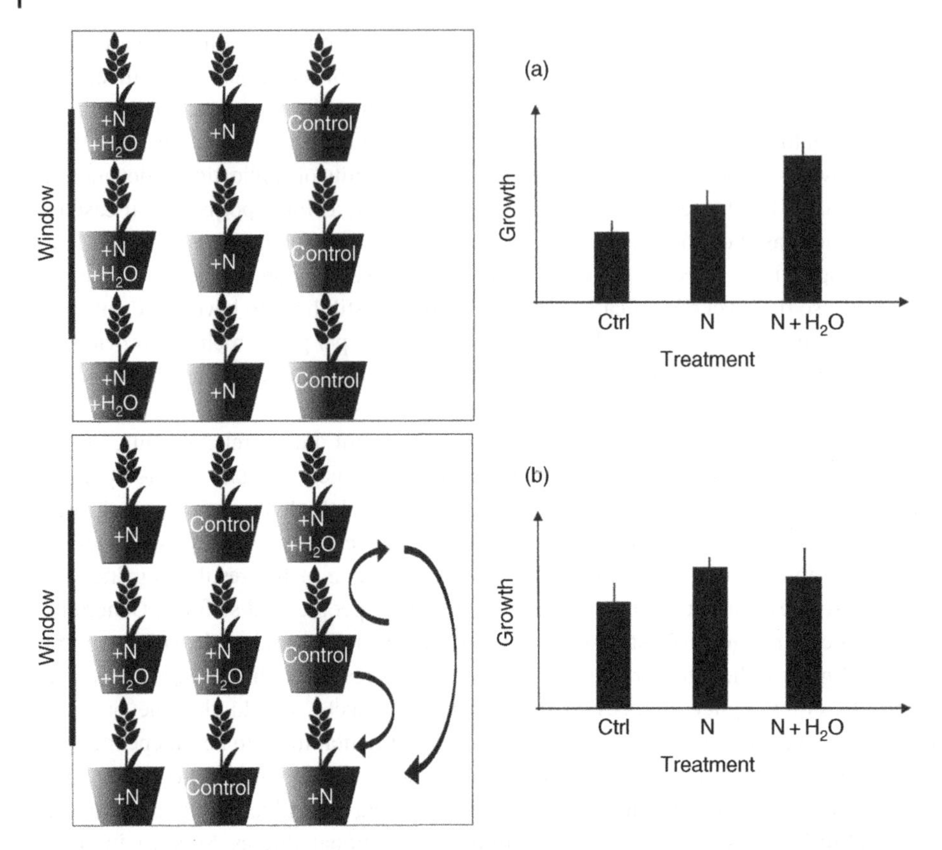

Figure 4.2 Two possibilities to arrange nine plant pots subjected to three different treatments in a hypothetical room with a window. In (a), the potential effect of the proximity to the window is confounded with the treatment effect, making it impossible to distinguish between the effect of the window, and that of the treatments (top right). The much better, randomised design in (b) prevents this: there does not seem to be a treatment effect.

Randomisation in scientific research is the random assignment of treatments to experimental units.

There would not be a problem with such a non-randomised placement of the experimental units if the conditions around the pots were perfectly homogenous. This is quite obviously not the case in our example because of the window, which we might associate with light and thus plant growth. If indeed the treatment plants that happen to be closest to the window grew more, then we might end up in a scenario as shown on the right-hand side of Figure 4.2a. At this point, the treatment effect is completely confounded by the possible but unknown effect of the window – we are unable to pinpoint the cause for more growth of the nitrogen plus water addition treatment plants. You might find this trivial and would intuitively tend to employ the randomised design shown in Figure 4.2b, which shows that in this case, the two treatments hardly have an effect on plant growth.

The phenomenon is far from trivial however, and often comes in disguise. It also extends past the position of the pots as discussed above and is of course not always as obvious as in our case. For example, if you receive a stack of pots that will be used for the experiment, and you start using the ones on top to assign to treatment 1, then to treatment 2, and eventually to the control, you might inadvertently introduce bias (systematic variation), perhaps because the top of the stack of pots was exposed to UV light more so than the bottom of the stack, which might have altered the chemical or physical properties of the pots. The same is true for the soil that is used to fill the pots and the seedlings or seeds that are used. Wherever you are assigning experimental units to treatments, this has to happen *randomly*. See Box 4.2 for a numerical illustration how failure to randomise can go wrong.

Box 4.2 Simulating the consequences of failure to randomise in R

Randomisation means we assign treatment units randomly to experimental units (e.g. plant pots to fertilised vs. unfertilised plants). Failure to do so can result in an increased risk of type I and type II errors. We will show this numerically here, again using a simple experiment on plants.

```
> set.seed(1) # use set seed to produce the same set of random numbers
> plantHeight <- rnorm(20, mean = 100, sd = 10) + 1:20
> treatment <- rep(c("ctrl", "treat"), each = 10)
```

The variable 'plantHeight' is the height of experimental plants that you received from a nursery. The variable 'treatment' is the treatment you want to subject those plants to (10 fertilised, 10 controls). Note that we simulate a slight trend in plant height, this could be because those at the back of the tray are slightly taller.

We now simply assign the first 10 plants to control pots, and the second 10 to treatment pots:

```
> d1 <- data.frame(plantHeight, treatment)
```

Having failed to randomise, we now test if the treatment has an effect on plant height, using a *t*-test (see Chapter 5):

```
> t.test(plantHeight ~ treatment, data = d1) # P-value = 0.03
```

It appears that the treatment had an effect, the *P*-value is significant. This is obviously wrong, and a mere consequence of us assigning the experimental units (pots) sequentially to subjects (plants), without randomising. The slight trend in original plant height got confounded with the treatment effect.

The correct way of planning and analysing this dataset would have been:

```
> d2 <- data.frame(plantHeight, treatment = sample(treatment))
> t.test(plantHeight ~ treatment, data = d2) # P-value = 0.54
```

Without proper randomisation, we cannot trust our *P*-values. You may receive a different result, depending on your random number generation, but on average, we will commit type I/II errors more often when we fail to randomise.

An even more subtle randomisation error is committed in the following example. Recall the grass fertiliser experiment from Box 4.1. We had four pots in each treatment, and five grasses per pot. It became clear that we have to work with the mean values per pot to avoid pseudoreplication:

```
> set.seed(2)
> treatment <- rep(c(0, 1), each = 4)
> biomass <- rnorm(8, mean = 100, sd = 5) # the mean values per pot
> d3 <- data.frame(treatment, biomass)
> t.test(biomass ~ treatment, data = d3) # P-value non-significant
```

So far, so good. The *t*-test is non-significant, which was expected, as we did not simulate a treatment effect. In an attempt to minimise the variance, a diligent technician homogenises the soil within each treatment. As a result, the mean biomass within each

(Continued)

Box 4.2 (Continued)

treatment remains the same, but the variation within treatment is reduced, say by a factor of 10 (we choose a high value to demonstrate the effect):

```
> x <- mean(d3$biomass[d3$treatment == 0])
> d3$biomass[d3$treatment == 0] <- x + (d3$biomass[d3$treatment == 0] - x)/10
> x <- mean(d3$biomass[d3$treatment == 1])
> d3$biomass[d3$treatment == 1] <- x + (d3$biomass[d3$treatment == 1] - x)/10
> t.test(biomass ~ treatment, data = d3) # P-value is significant!
```

Note that as a result of the *treatment-wise* homogenisation of the soil, the *t*-test becomes significant despite there being no simulated treatment effect. This happens because we did not randomise correctly, we minimised the variation *within* treatments separately instead of across all treatments.

4.4 Randomisation in R

How do we randomise, i.e. assign experimental units to treatments randomly? One might be tempted to just 'do it randomly', or deliberately make sure that the assignment looks random to our eye. This technique is sometimes termed 'haphazard' allocation of experimental units to treatments, but it is not the method of choice. Humans are shockingly bad random number generators. If humans are told to assign something 'randomly', then it can be shown that on average, we are not performing well. Where and when possible, we want to use proper random number generators in R, or in more complicated cases, even a dedicated R package. To randomly assign the treatments to the positions in our example, we can use this code, for example:

```
> treatments <- rep(c('N', 'N_H2O', 'ctrl'), each = 3)
> data.frame(treat = sample(treatments))
   treat
1      N
2      N
3   ctrl
4  N_H2O
5  N_H2O
6   ctrl
7      N
8  N_H2O
9   ctrl
```

The function `sample` shuffles the treatments randomly, and the row index (from 1 to 9) of the data frame represents the position of the pots in the experimental chamber. Over the duration of the experiment, you can and should repeat the randomisation procedure in order to reduce bias.

Things can however become more complex if you have hierarchical or split-plot designs. For example, if you include a treatment that cannot be separated spatially (e.g. if you have a whole room set to a given temperature), you should *not* continuously randomise *within* one treatment. This is because it reduces the variation within treatments, which in turn artificially increases the chances to find differences between treatments (see Box 4.2). This can in turn inflate type I errors, and it is a good idea to get advice if you are unsure how to randomise in more complex study designs. The R package `randomizeR` can help with this too.

The function `sample` can also help choose subjects from a pool, for example when you need to choose 10 samples out of a selection of 100. Because you might not be able (or want to) sample a subject twice, the argument `replace` should be set to `FALSE` (which it is by default):

```
> sample(1:20, size = 10) # sample without replacement
 [1]  5 15 13 11  3 16 10  8  6  4
> sample(1:20, size = 10, replace = TRUE) # sample with replacement
 [1] 13  6  6 17  4  1 20  4  6 17
```

In many cases, a simple random number generator can also do the job. You can, for example number your experimental units and then ask 'which unit do I assign next?' for which the answer could be

```
> round(runif(1, min = 0, max = 20.49))
[1] 18
```

Or in a spatial design as shown below, you could get random x and y coordinates using:

```
x_coord <- runif(10, min = 0, max = 10)
y_coord <- runif(10, min = 0, max = 10)
```

As we will see now, spatial designs can be slightly more complex, and different approaches are possible.

4.5 Spatial Replication and Randomisation in Observational Studies

If a research question involves spatial sampling (e.g. water samples from a lake, soil samples from an island, insect traps placed in a forest), the decision of *where* exactly we take our samples becomes critical. As we have seen, sampling 'haphazardly' by deliberately choosing seemingly independent sites is not a great option. It should be noted however that this is sometimes the only practical option, particularly if there are other constraints (for example safety concerns). A first option is systematic sampling, where an equally spaced grid is overlaid with the area that needs to be sampled. If you are sampling a gradient, we can simply allow equal distances between our samples. This avoids spatial clustering and ensures independence to some degree. Systematic sampling is illustrated in Figure 4.3a, using the example of an island. An alternative to systematic sampling is random sampling (Figure 4.3b). To select your sampling sites, you can use `runif` or `sample` to determine your x and y coordinates as shown above. This is a bit easier if you have a square but can be applied to an irregularly shaped island as shown in Figure 4.3b, simply by ignoring sites that fall on water. Figure 4.3c combines the two approaches (stratified random sampling). This can be useful to avoid undersampled or oversampled areas (the southern side of the island seems under-sampled in Figure 4.3b). It also ensures that we have samples from certain areas that are of interest (note that the surface areas of the strata do not have to be equal). There are two arguments in favour of random or stratified random sampling rather than systematic sampling. First, edge effects might manifest in systematic sampling. This can be seen from Figure 4.3a, where the line of sample sites coincides with the coastline of the

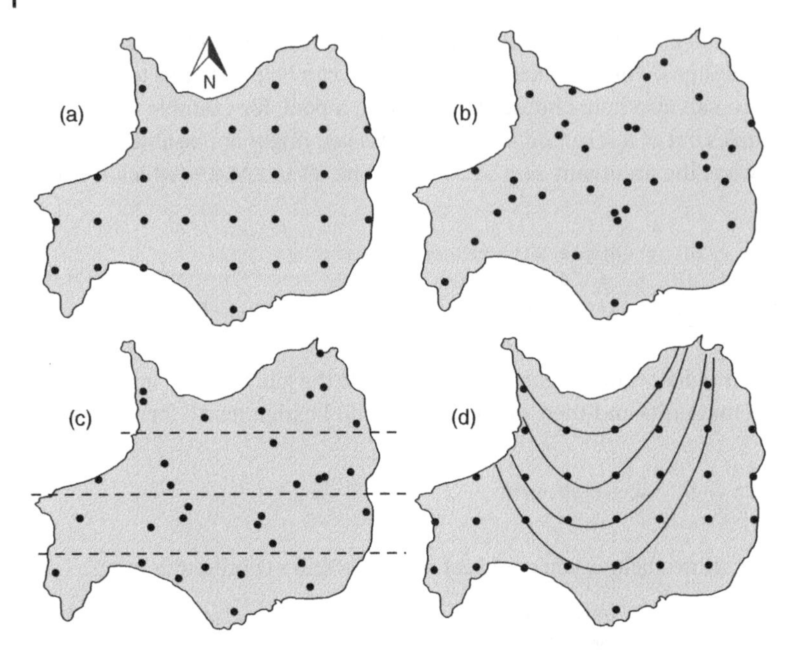

Figure 4.3 Spatial sampling protocols on maps of the Isle of Gork using 32 sampling sites. (a) Systematic sampling, (b) random sampling, (c) stratified random sampling. (d) Visualises the problem with systematic sampling if there are underlying patterns in the soil (black lines) that match the periodicity of the sampling grid.

island on the eastern side. Second, if there are underlying systematic patterns that share the periodicity or spacing of your grid, your samples may no longer be independent. This is illustrated by the black lines in Figure 4.3d, which could be bedrock formations, ground water, vegetation patterns, etc. Such effects are subtle and need to be assessed for every individual scenario.

The examples in this chapter should sharpen our statistical sense for how important the concepts of replication, randomisation, and independence are. These should be used both not only to plan your own scientific studies but also to scrutinise the work of others. Replication as *the* tool to separate signal from noise is simply indispensable, but without properly characterising at which level the experimental units are independent, we risk pseudoreplication. This, as well as failure to randomise can lead to erroneous statistical conclusions (inflated type I and II errors, see Chapter 5). Note that the examples used in this chapter may appear unrealistically simple, and things will likely not be as obvious in the real world.

5

Two-Sample and One-Sample Tests

One- and two-sample tests were the early cornerstones in statistical data analysis. Nowadays, however, our experimental designs tend to be more complex than what these simple tests can handle. Nonetheless, these tests, the famous Student's t-test in particular, provide excellent entry points into statistical modelling, since the underlying principle can be understood quite easily and they get you into the swing of understanding hypothesis testing and the interpretation of test statistics such as t-, z-, F-, χ^2-values, etc.

5.1 The *t*-Statistic

The t-statistic boils down to a simple signal-to-noise ratio where the difference between two group means (the signal or effect) is normalized (divided) by the pooled standard deviation of the two groups (the noise, Figure 5.3).

$$t = \frac{\text{Signal}}{\text{Noise}} = \frac{\text{Difference between means}}{\text{Pooled standard deviation}} = \frac{\bar{x}_1 - \bar{x}_2}{\sqrt{\frac{s_1^2}{n_1} + \frac{s_2^2}{n_2}}} \tag{5.1}$$

This test-statistic follows a symmetric, bell-shaped distribution similar to the normal distribution but with longer tails that is called Student's t-distribution. The number of degrees of freedom $(n_1 + n_2 - 2)$ is the shape defining parameter of the t-distribution (Figure 5.1).

5.2 Two Sample Tests: Comparing Two Groups

5.2.1 Student's *t*-Test

Student's t-test is applied in situations where we have a continuous response variable and two groups (two samples) to compare. The two groups represent a binary explanatory variable such as a crop fertiliser treatment, where we have a control group without fertiliser addition and a treatment group receiving the fertiliser product. The test assumes:

- that the data approximately follows a ***normal distribution*** in each group
- ***homogeneity of variances***, i.e. the variance of the response variable should be equal in each group

R-ticulate: A Beginner's Guide to Data Analysis for Natural Scientists, First Edition.
Martin Bader and Sebastian Leuzinger.

Companion website: www.wiley.com/go/Bader

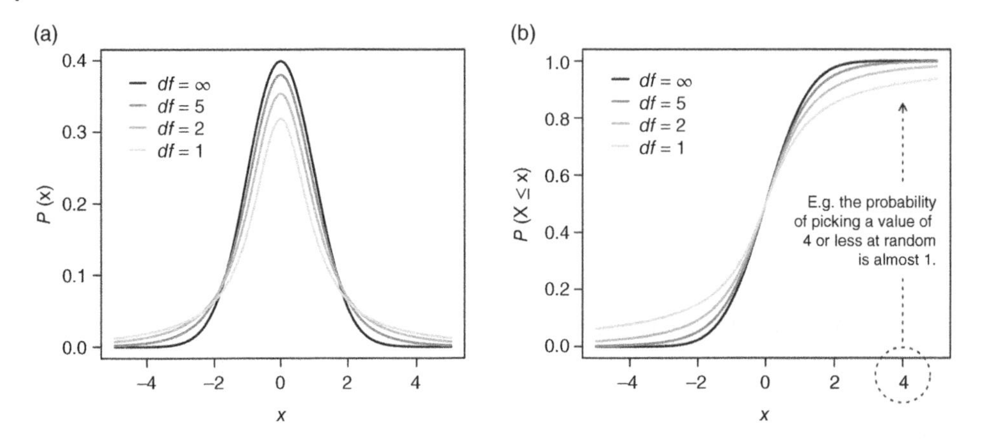

Figure 5.1 (a) Probability density function (PDF) and (b) cumulative distribution function (CDF) of Student's t-distribution. The probabilities given by the PDF sum up to 1 (100%) represented by the area under the curve. The PDF can thus be used to determine the probability of the random variable X to fall within a particular range of values (continuous probability functions are defined for an infinite number of points over a continuous interval and thus the probability at a single point is always zero). The CDF gives the probability that a real-valued random variable X will take on a value less than or equal to x (the theoretical quantiles).

- ***independence of observations***, i.e. within each group, samples are independently and randomly drawn from the population. Consequently, each observation should belong to only one group and there should be no relationship between the observations within a group.

5.2.1.1 Testing for Normality

We can use quantile-quantile plots or histograms as visual tools to evaluate whether the data is normally distributed (Figure 5.2), or choose a formal statistical procedure in the form of one of the many available normality tests, of which perhaps the most popular are:

- *Shapiro–Wilk* test
- *Kolmogorov–Smirnov* (K–S) test
- *Anderson–Darling* test (a modified version of the K–S test)
- *Cramér–von Mises* test (a more powerful version of the K–S test)

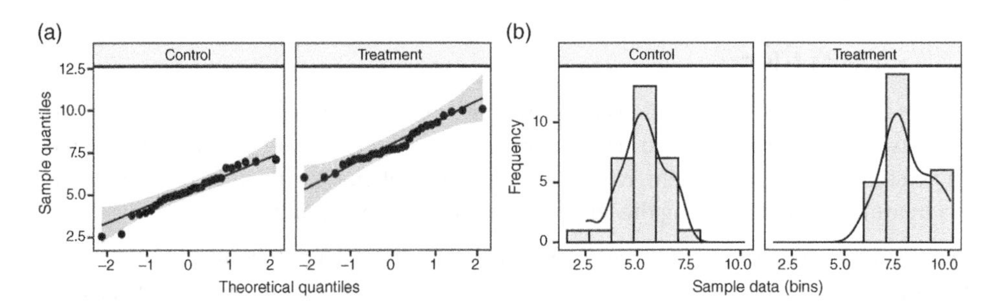

Figure 5.2 Quantile–quantile plots (**Q-Q plot**, a): a visual tool to assess normality by plotting a vector of random variables against the quantiles (values) of a theoretical normal distribution. If the data points are close to the quantile–quantile line, we can assume that they approximately follow a normal distribution. Histograms (b) visualize the distribution of a continuous random variable, here overplotted with a kernel density estimate of the probability density of the data.

Additional normality tests can be found in the R package `nortest`.

We can either form subsets for each group and apply the graphical normality checks and formal normality tests separately or do it all in one-go using the functionality given in the *tidyverse* collection of R packages which were designed for efficient data wrangling.

```
## Load required packages
> library(tidyverse)
> library(broom)
> library(rstatix)
> library(ggpubr)

## Dummy t-test data
> set.seed(2)
> dat <- data.frame(response = round(c(rnorm(n = 30, mean = 5),
                                      rnorm(n = 30, mean = 8)), digits = 2),
                    treat = rep(c("control", "treatment"), each = 30))

> head(dat, n = 3)
  response   treat
1     4.10 control
2     5.18 control
3     6.59 control
## Group subsets
> ctrl <- dat[dat$treat == "control", "response"]
> trt <- dat[dat$treat == "treatment", "response"]

## Quantile-quantile plots
## Control group
> qqnorm(ctrl)
> qqline(ctrl)
## Treatment group
> qqnorm(trt)
> qqline(trt)
## Histogram
> hist(ctrl, breaks = 6)
> hist(trt, breaks = 5)
## Quantile-quantile plot
> ggqqplot(dat, x = "response", facet.by = "treat")
## Density plot
> gghistogram(data = dat, x = "response", facet.by = "treat", fill = "lightgray",
          bins = 8, add_density = T, xlab = "Response",
          ylab = "Probability density")
```

The null hypothesis underlying the normality tests states that the distribution of our sample data is not significantly different from a normal distribution and therefore we expect the resulting *P*-value to be *larger* than 0.05, if our data really follows a normal distribution. This is in contrast to the majority of analysis scenarios down the road, where we often expect a significant effect and thus a *P*-value smaller than 0.05 prompting us to reject the null hypothesis and accept the alternative hypothesis (there is an effect). Consequently, a significant outcome of a normality test indicates that the tested data follows a distribution other than the normal distribution.

The ***Shapiro–Wilk*** test is perhaps the best known normality test. The test is performed using the `shapiro.test` function, which only requires a single input vector holding the data to be tested for normality.

```
## Shapiro-Wilk normality test
> shapiro.test(ctrl)

> Shapiro-Wilk normality test

data:  ctrl
W = 0.96476, p-value = 0.4072

> shapiro.test(trt)

Shapiro-Wilk normality test

data:  trt
W = 0.95121, p-value = 0.1822
```

Similar to the graphical exploration earlier, we can use a *tidyverse* approach and for the Shapiro–Wilk test there is even the `shapiro_test` wrapper around the base function `shapiro.test` making it easy to use when coding with grouped data and pipes (recognisable by the pipe operator `%>%`).

```
## Shapiro-Wilk normality test
> group_by(.data = dat, treat) %>% shapiro_test(response)
# A tibble: 2 x 4
  treat     variable statistic     p
  <chr>     <chr>        <dbl> <dbl>
1 control   response     0.965 0.407
2 treatment response     0.951 0.182
```

The ***Kolmogorov–Smirnov*** (**K–S**) test compares the distribution of a set of random variables to a user-chosen reference distribution (one-sample K–S test). Since the test compares the empirical distribution function of the sample to the cumulative distribution function (CDF) of a reference distribution and because the CDFs in R all start with 'p' (e.g. `pnorm`, `ppois`, `pgamma`, etc.), we need to specify 'pnorm' in the `ks.test` function. Alternatively, we can use the K–S test to compare two sets of random variables to test whether they share the same distribution (two-sample K–S test).

```
## Kolomogorov-Smirnov test
## One-sample test comparing the input data to a
## theoretical normal distribution given the same
## mean and standard deviation

## Old school coding
> ks.test(x = ctrl, y = "pnorm", mean = mean(ctrl), sd = sd(ctrl))

One-sample Kolmogorov-Smirnov test

data:  ctrl
D = 0.078301, p-value = 0.9929
alternative hypothesis: two-sided

> ks.test(x = trt, y = "pnorm", mean = mean(trt), sd = sd(trt))

One-sample Kolmogorov-Smirnov test

data:  trt
D = 0.15244, p-value = 0.4885
alternative hypothesis: two-sided

> ks.test(ctrl, trt)
```

```
Two-sample Kolmogorov-Smirnov test

data:  ctrl and trt
D = 0.83333, p-value = 1.792e-09
alternative hypothesis: two-sided

## tidyverse coding
> group_by(.data = dat, treat) %>%
  group_modify(.f = ~ tidy(ks.test(x =.x$response, y = "pnorm",
                                   mean = mean(.x$response),
                                   sd = sd(.x$response))))

# A tibble: 2 x 5
# Groups:   treat [2]
  treat   statistic p.value method               alternative
  <chr>       <dbl>   <dbl> <chr>                <chr>
1 control    0.0783   0.993 One-sample Kolmogor~ two-sided
2 treatm~    0.152    0.488 One-sample Kolmogor~ two-sided
```

The ***Anderson–Darling*** test is a modified, more powerful version of the K–S test giving more weight to the tails, which can be performed using the function `ad.test` (R package `nortest`).

```
## Anderson-Darling test
> library(nortest)

## Old school coding
> ad.test(ctrl)

Anderson-Darling normality test

data:  ctrl
A = 0.2676, p-value = 0.6612

> ad.test(trt)

Anderson-Darling normality test

data:  trt
A = 0.5375, p-value = 0.1547

## tidyverse coding
> group_by(.data = dat, treat) %>%
  group_modify(.f = ~ tidy(ad.test(.x$response)))

# A tibble: 2 x 4
# Groups:   treat [2]
  treat     statistic p.value method
  <chr>         <dbl>   <dbl> <chr>
1 control       0.268   0.661 Anderson-Darling normality te~
2 treatment     0.537   0.155 Anderson-Darling normality te~
```

Similarly, the ***Cramér–von Mises*** test is a more powerful version of the K–S test that can be run with the command `cvm.test` (R package *nortest*).

```
## Cramér-von Mises test
## Old school coding
> cvm.test(ctrl)

Cramér-von Mises normality test

data:  ctrl
W = 0.029973, p-value = 0.841
```

```
> cvm.test(trt)

Cramér-von Mises normality test

data:  trt
W = 0.099592, p-value = 0.1081

## tidyverse coding
> group_by(.data = dat, treat) %>%
  group_modify(.f = ~ tidy(cvm.test(.x$response)))
# A tibble: 2 x 4
# Groups:   treat [2]
  treat     statistic p.value method
  <chr>         <dbl>   <dbl> <chr>
1 control      0.0300   0.841 Cramér-von Mises normality te~
2 treatment    0.0996   0.108 Cramér-von Mises normality te~
```

Apart from normally distributed data, Student's t-test assumes equal variances among the two groups to be compared. This so-called *variance homogeneity* criterion can be formally tested using a range of variance tests such as Levene's test (`levene_test`, R package *rstatix*), Barthlett's test (`bartlett.test`, built-in), or the Fligner–Killeen test, (`fligner.test`, built-in) to name but a few. Similar to the normality tests, the null hypothesis of variance homogeneity tests assumes no difference in group variances and thus a statistically significant test result indicates unequal variances among groups.

```
## Checking for equal variances
> levene_test(data = dat, formula = response ~ treat)

# A tibble: 1 x 4
    df1   df2 statistic     p
  <int> <int>     <dbl> <dbl>
1     1    58  0.000691 0.979
```

In case of equal group variances, the pooled variance is used as overall variance estimate. However, if this assumption is violated, then the variance needs to be estimated separately for each group, which involves a modification to the degrees of freedom, known as *Welch approximation*. Since Welch's t-test version is superior to Student's t-test when sample sizes and variances are unequal between groups and yields the same result when sample sizes and variances are equal, the built-in `t.test` function uses the Welch approximation by default, so testing for equal variances in connection with a t-test is actually redundant. This default setting is implemented by the `var.equal = FALSE` argument (the default setting). When set to `var.equal = TRUE`, the `t.test` function assumes equal group variances and computes the original Student's t-test.

As we have seen earlier, the t-statistic is easily computed as the difference between group means divided by the pooled standard deviation (as a measure of spread), which can be regarded as a simple signal-to-noise ratio (signal: difference between means, noise: pooled standard deviation; Figure 5.3). The test procedure then queries the probability density function (PDF) underlying the t-distribution to obtain an answer to the question:

How likely is it to obtain a t-value as extreme as the observed one just by chance?[1]

We then compare the resulting probability, the so-called P-value, to the probability of the critical t-value corresponding to our preset significance level of 5% (by convention denoted

1 Note, the use of the term 'extreme' instead of 'large' to account for the fact that the sign of the test statistic can be negative or positive. One could rephrase the question as: How likely is it to obtain a t-value as large as the absolute value of the observed one, just by chance?

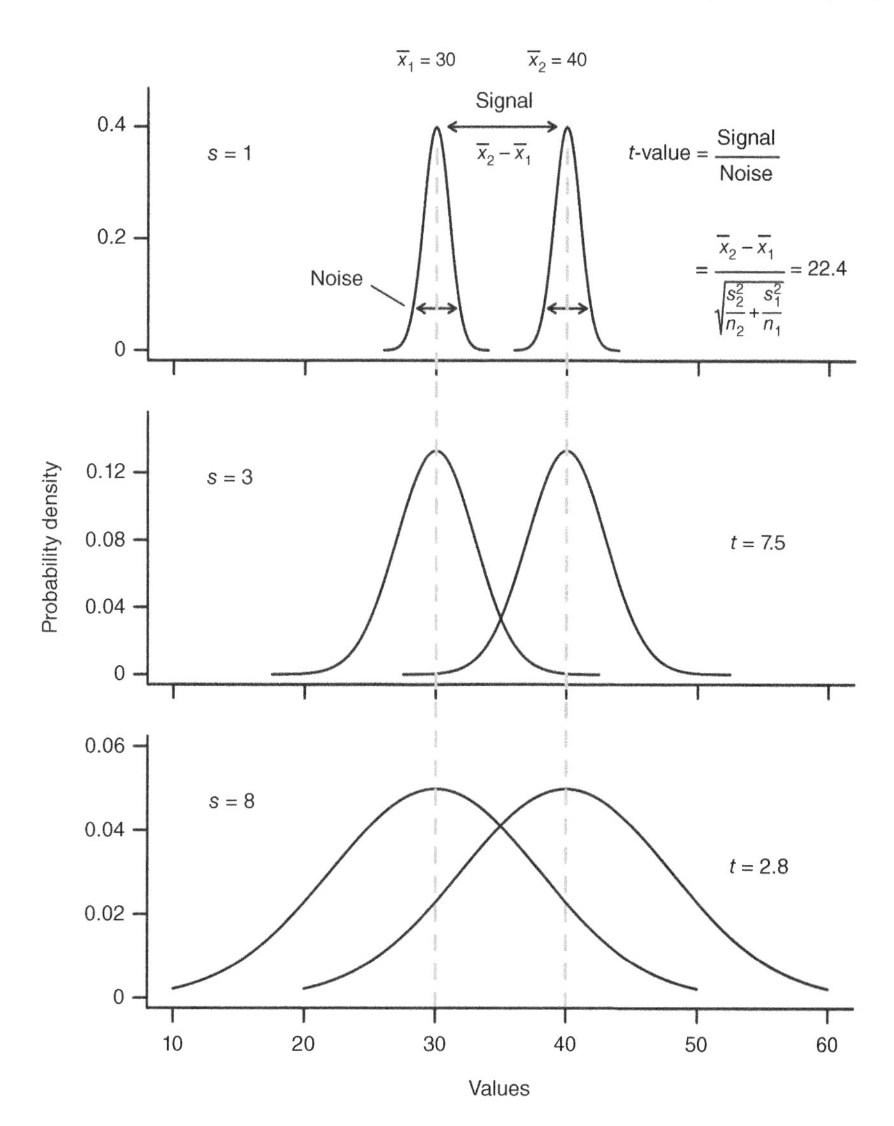

Figure 5.3 Three different t-test scenarios with identical means but increasing amount of spread (greater standard deviation) from top to bottom. Note the reduction of the t-statistic in response to the increasing standard deviation (increase in noise).

by $\alpha = 0.05$). If our P-value is smaller than 0.05, we reject the null hypothesis of no difference among the two groups in favour of the alternative hypothesis and conclude that there is a statistically significant difference between the two groups.

The t-test can be specified using a formula interface (in our view this is the preferred option since more sophisticated models all rely on it) or using a simple x, y notation.

```
## Welch's t-test, formula interface
> t.test(response ~ treat, data = dat)

Welch Two Sample t-test

data:  response by treat
t = -9.1328, df = 57.973, p-value = 8.045e-13
```

```
alternative hypothesis: true difference in means is not equal to 0
95 percent confidence interval:
 -3.336897 -2.137103
sample estimates:
  mean in group control mean in group treatment
                  5.228                   7.965

## Traditional interface using 'x, y' notation and the previously created subsets
> t.test(x = ctrl, y = trt)


Welch Two Sample t-test

data:  ctrl and trt
t = -9.1328, df = 57.973, p-value = 8.045e-13
alternative hypothesis: true difference in means is not equal to 0
95 percent confidence interval:
 -3.336897 -2.137103
sample estimates:
mean of x mean of y
    5.228     7.965
```

5.2.1.2 What to Write in a Report or Paper and How to Visualise the Results of a *t*-Test

We can wrap up the results in one concise sentence like this: The treatment caused a statistically significant increase in the response by 52% compared to the control ($t = -9.13$, $df = 57.97$, $P < 0.001$).

The results are best displayed as bar- or boxplots (Figure 5.4). First, we need to aggregate our data, i.e. compute the group means and the associated standard errors.

```
## Aggregate the data for plotting
> dat2 <- group_by(dat, treat) %>% summarise(response_mean = mean(response),
                                            pos = mean(response) + se(response),
                                            neg = mean(response) - se(response))
> bp <- barplot(dat2$response_mean, las = 1, ylim = c(0, 10),
              col = c("white", "grey"))
> bp # midpoint of the bars

> arrows(x0 = bp, y0 = dat2$neg, x1 = bp, y1 = dat2$pos, length = 0.05,
       angle = 90, code = 3)
> text(x = bp, y = dat2$pos, labels = c("a", "b"), pos = 3)
> mtext(text = "Response", side = 2, line = 2.5, cex = 0.9)
> mtext(text = c("C", "T"), side = 1, line = 0.5,
      cex = 0.9, at = bp)
```

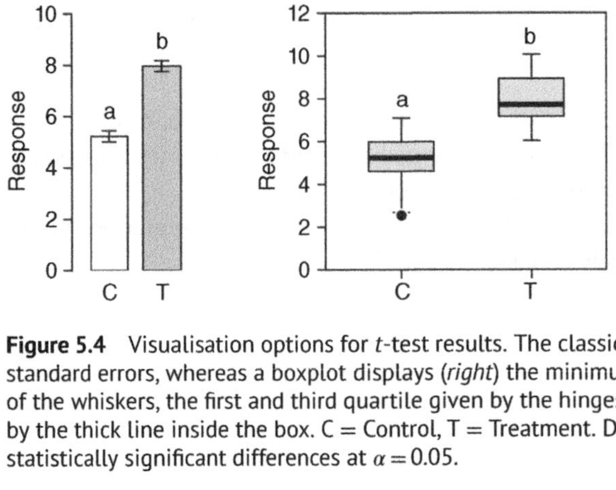

Figure 5.4 Visualisation options for *t*-test results. The classical barplot (*left*) shows the means ± standard errors, whereas a boxplot displays (*right*) the minimum and maximum indicated by the end of the whiskers, the first and third quartile given by the hinges of the box, and the median denoted by the thick line inside the box. C = Control, T = Treatment. Different lower-case letters indicate statistically significant differences at $\alpha = 0.05$.

5.2.1.3 Two-Tailed vs. One-Tailed *t*-Tests

The *t*-test allows us to specify the alternative hypothesis via the `alternative` argument, whose default is 'two-sided', translating into the alternative hypothesis: 'the difference between group means is not zero'. This means that we make no *a priori* assumption as to whether the mean of group one ($\overline{x}_1$) is smaller or greater than the mean of group two ($\overline{x}_2$). So, the resulting *t*-value can be positive ($\overline{x}_1 > \overline{x}_2$) or negative ($\overline{x}_1 < \overline{x}_2$), and to account for this, the 5% *type I error* probability is divided between the two tails of the distribution. Setting the `alternative` argument to 'less', results in a one-tailed test assuming that the mean of the first group is smaller than the mean of the second group and *vice versa* when 'greater' is specified (Figure 5.5). One-tailed *t*-tests should only be applied if we have very strong reason to believe that the difference can only go in one particular direction. Shifting the 5% error probability all in one tail reduces the critical *t*-value and therefore one-tailed tests have more statistical power to detect an effect in one particular direction than a two-tailed test applied to the same data.

```
## One-sided t-test

# H0: first mean is smaller than the second
> t.test(response ~ treat, data = dat, alternative = "less")

Welch Two Sample t-test

data:  response by treat
t = -9.1328, df = 57.973, p-value = 4.022e-13
alternative hypothesis: true difference in means is less than 0
95 percent confidence interval:
      -Inf -2.236052
```

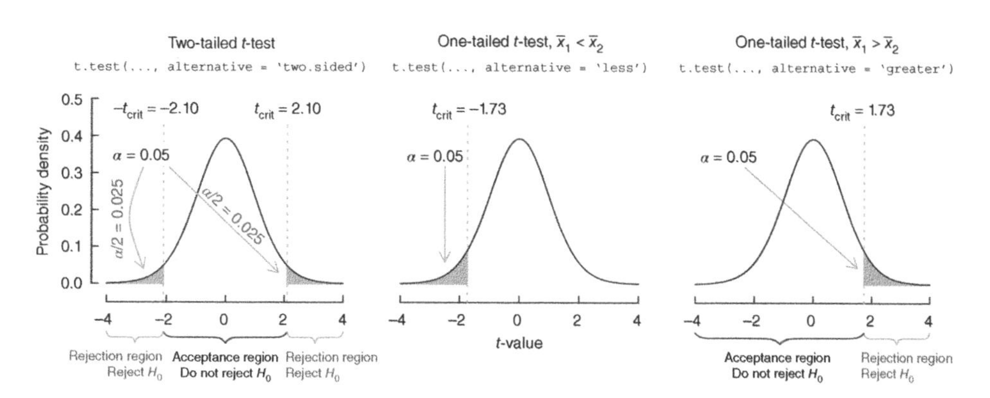

Figure 5.5 Two-tailed vs. one-tailed *t*-tests. By default the `t.test` function runs as a two-tailed test meaning that we make no assumptions as to which group mean is larger. Consequently, the *t*-statistic may be positive ($\overline{x}_1 > \overline{x}_2$) or negative ($\overline{x}_1 < \overline{x}_2$), i.e. it may sit in the left or right tail of the distribution and in order to take this into account, we need to split our 5% *type I error* probability, given by the significance level ($\alpha = 0.05$), in two and allocate one half to the left tail and the other half to the right tail (left panel). If we have strong reason to believe that the mean of group 2 can only be larger than that of group 1 ($\overline{x}_1 < \overline{x}_2$), then we can shift the whole 5% error probability into the left tail resulting in a one-tailed *t*-test (middle panel). In the opposite case, when we strongly assume that $\overline{x}_1 > \overline{x}_2$, we shift the whole 5% error probability into the right-hand tail (right panel). Note the lower critical *t*-values associated with shifting the whole 5% error probability into one tail, which makes a one-tailed *t*-test more powerful than a two-tailed one for detecting changes in one particular direction. The shape of the *t*-distribution and the critical *t*-values (t_{crit}) are based on a hypothetical example with 10 observations per group and thus $n_1 + n_2 - 2 = 18$ degrees of freedom.

```
sample estimates:
  mean in group control mean in group treatment
                  5.228                   7.965

# H0: first mean is larger than the second
> t.test(response ~ treat, data = dat, alternative = "greater")


Welch Two Sample t-test

data:  response by treat
t = -9.1328, df = 57.973, p-value = 1
alternative hypothesis: true difference in means is greater than 0
95 percent confidence interval:
 -3.237948       Inf
sample estimates:
  mean in group control mean in group treatment
                  5.228                   7.965
```

In 'before and after' experiments, e.g. when the efficacy of a treatment is tested on the same subjects, the observations are not independent, which commonly results in less variation compared to data derived from studies contrasting independent groups. In those situations, a paired t-test is required to account for the dependence among observations and this test mode can be implemented by setting `paired = TRUE`. Consider the following small dataset on depression scale measurements in nine patients taken at the first (i) and second (ii) visit after initiation of a therapy (administration of a tranquiliser).

```
> depression <- data.frame(score = c(1.83,  0.50,  1.62,  2.48, 1.68, 1.88, 1.55,
                                    3.06, 1.30, 0.83, 0.61, 0.57, 2.03, 1.06,
                                    1.29, 1.06, 3.14, 1.29),
                         visit = rep(1:2, each = 9), patient = 1:9)
> t.test(score ~ visit, data = depression, paired = T)

Paired t-test

data:  score by visit
t = 3.094, df = 8, p-value = 0.0148
alternative hypothesis: true difference in means is not equal to 0
95 percent confidence interval:
 0.1137571 0.7795763
sample estimates:
mean of the differences
              0.4466667
```

So, what is the magic behind a paired t-test?

In order to generate an independent dataset, the paired t-test computes the *differences* between before and after observations of each patient. With this little trick we get rid of the dependency. In the original dataset, we had two observations per patient (before and after) but using the differences leaves us with only one observation per patient and makes this an independent dataset. We can now test the before-after differences against the null hypothesis that the mean difference is zero. This sort of analysis is done using a one-sample t-test and this is what goes on behind the scenes of a paired t-test: it runs a one-sample t-test on the differences. We will learn about one-sample t-tests in Section 5.3.

5.2.2 Rank-Based Two-Sample Tests

The ***Wilcoxon Rank Sum test*** is a non-parametric alternative to the *t*-test. Non-parametric tests do not assume an underlying probability distribution of the sample data, meaning that our data does not need to be normally distributed. The test procedure relies on ranks rather than on the actual values and thus makes it very robust against outliers (which do have a strong influence on the *t*-test though).

```
## Robustness of ranks against outliers
> a <- c(2, 3, 5); b <- c(3, 6, 8) # two samples, no outliers
> x <- c(0, 3, 4); y <- c(3, 23, 700) # two samples, one with an outlier

> rank(c(a, b))
[1] 1.0 2.5 4.0 2.5 5.0 6.0

> rank(c(x, y))
[1] 1.0 2.5 4.0 2.5 5.0 6.0 # same ranking despite the outlier
```

The test is implemented by the `wilcox.test` function, which also features a `paired` argument, allowing the analysis of paired designs (dependent observations). In the unpaired case (independent samples), we talk about the *Wilcoxon rank sum* test, which is equivalent to the *Mann–Whitney* test (a.k.a. *U*-test). In the paired case (dependent samples), we refer to this procedure as the *Wilcoxon signed rank* test.

```
## Formula notation
> wilcox.test(response ~ treat, data = dat)

Wilcoxon rank sum test with continuity correction

data:  response by treat
W = 23.5, p-value = 3.005e-10
alternative hypothesis: true location shift is not equal to 0

## Traditional interface using 'x, y' notation
> wilcox.test(x = ctrl, y = trt)

Wilcoxon rank sum test with continuity correction

data:  ctrl and trt
W = 23.5, p-value = 3.005e-10
alternative hypothesis: true location shift is not equal to 0
```

In the same manner as the *t*-test, the Wilcoxon test can be run as a one-tailed test or a paired test, e.g. the analysis of the previously used `depression` dataset.

```
> wilcox.test(score ~ visit, data = depression, paired = T)

Wilcoxon signed rank exact test

data:  score by visit
V = 40, p-value = 0.03906
alternative hypothesis: true location shift is not equal to 0
```

5.3 One-Sample Tests

At first glance, a one-sample test seems paradoxical but sometimes we simply want to compare a sample mean with a known value, e.g. when we would like to test whether a fold-change is significantly different from 1 or when we want to compare the mean of a randomly chosen sample to a known population mean. In those cases, a one-sample t-test or a one-sample *Wilcoxon* test can be applied by providing the observations of the sample and a single value for the population mean instead of the observations from a second group. Imagine a crop scientist who was interested in determining whether priming of wheat seeds with silica resulted in a larger average height of wheat seedlings than the standard height of 10.5 cm. The researcher treated a random sample of $n = 33$ seeds with silica nanoparticles prior to sowing and subsequently obtained the seedling heights in cm given in the `seeds` vector as defined in the following code.

```
## Dummy data
> seeds <- c(11.5, 11.8, 15.7, 16.1, 14.1, 10.5, 15.2, 19.0, 12.8, 12.4, 19.2,
             13.5, 16.5, 13.5, 14.4, 16.7, 10.9, 13.0, 15.1, 17.1, 13.3, 12.4,
             8.5, 14.3, 12.9, 11.1, 15.0, 13.3, 15.8, 13.5, 9.3, 12.2, 10.3)
> mean(seeds)
[1] 13.66364
## The mean is 13.7 cm but is this a significant increase compared to
## the 10.5 cm standard height?
> t.test(seeds, mu = 10.5)

One Sample t-test

data:  seeds
t = 7.1449, df = 32, p-value = 4.14e-08
alternative hypothesis: true mean is not equal to 10.5
95 percent confidence interval:
 12.76172 14.56556
sample estimates:
mean of x
 13.66364

> wilcox.test(seeds, mu = 10.5)

Wilcoxon signed rank test with continuity correction

data:  seeds
V = 512, p-value = 3.682e-06
alternative hypothesis: true location is not equal to 10.5
```

Both one-sample tests give us a significant result indicating that silica seed priming significantly increases seedling height by 30% (`mean(seeds)/10.5 - 1`).

At this point, we will revisit the ‘depression’ example to unveil the inner workings of a paired t-test. Let us calculate the difference in depression scores between the first and the second visit and analyse it with a one-sample t-test.

```
> difference <- depression[depression$visit == "1", "score"] -
                depression[depression$visit == "2", "score"]
> t.test(difference, mu = 0)

One Sample t-test

data:  difference
```

```
t = 3.094, df = 8, p-value = 0.0148
alternative hypothesis: true mean is not equal to 0
95 percent confidence interval:
 0.1137571 0.7795763
sample estimates:
mean of x
0.4466667
```

When we compare the outcome of the one-sample *t*-test applied to the differences in depression scores to a paired *t*-test using the original data, we can see that they yield identical results, which shows that a paired *t*-test boils down to a one-sample test with the null hypothesis that the difference between the mean of the sample and a reference mean is zero.

```
> t.test(score ~ visit, data = depression, paired = T)

Paired t-test

data:  score by visit
t = 3.094, df = 8, p-value = 0.0148
alternative hypothesis: true difference in means is not equal to 0
95 percent confidence interval:
 0.1137571 0.7795763
sample estimates:
mean of the differences
              0.4466667
```

5.4 Power Analyses and Sample Size Determination

The power of a hypothesis test is the probability of correctly rejecting the null hypothesis (H_0) when the alternative hypothesis (H_A) is true (Chapter 3, Figure 3.9). In other words: the power indicates the probability of avoiding a type II error (β; a *false negative*, failing to reject a false null hypothesis) and is thus defined as $1 - \beta$ (hatched area in Chapter 3, Figure 3.9). Like every probability, the power of a test ranges from 0 to 1 and with increasing statistical power, the probability of committing a type II error decreases. There are four factors affecting the power of a statistical test:

- The **sample size** n.
- The **significance level** (α). The lower α, the lower the power of the test. Reducing α (e.g. from 0.05 to 0.01) widens the region of acceptance and as a result, we are less likely to reject H_0 even when it is false, so we are more likely to commit a type II error (β).
- The **effect size**. The bigger the effect size, the greater the power of the test.
- the **standard deviation** of the samples. The smaller the standard deviation, the greater the power of the test.

Before you start an experiment or an observational study, the fundamental question of how many samples to take will arise. Because of the aforementioned dependencies, the answer to this question requires an estimate of (i) the variation in your population and (ii) the expected difference between treatments or groups (effect size). To get estimates for these metrics, you need to perform a literature search or conduct a pilot study. Determining reasonable estimates is a good time investment: think of a situation where an expensive experiment is conducted and afterwards it is found that given the variability in the data, the expected difference could never have been detected. For example, if your expected difference is 1 (arbitrary units), the mean standard deviation in your samples is 3 and your sample size 10, then your power is as low as 10%! In other words, you only have a 10% chance of detecting a potential difference of 1 between your groups! In this case, you would have to increase your sample size dramatically to about 150 to reach a power of around 80%.

We can use the versatile `power.t.test` function to juggle with sample size, significance level, power, standard deviation, and the expected difference between groups to help guide our experimental design. Whatever single parameter we do not specify in this function will be determined from the others and returned in the output. For example, if we provide the sample size, the estimated difference between means, the standard deviation and the significance level, then the function returns the power of the resulting *t*-test. By convention, we aim at a target power of 80% (0.8), so we can plug this value into the function along with all other required arguments apart from the one parameter of interest we would like to get an estimate for. This is great for sample size determination which can be derived by omitting the `n` argument. By default the `power.t.test` function assumes a significance level of 0.05 (argument `sig.level`), a two-sample test (arg. `type`) with a `two.sided` alternative hypothesis (arg. `alternative`) and therefore these arguments need not be specified unless we wish to change them.

Let us start with a simple power calculation to determine whether it is worthwhile conducting a planned experiment. Say we expect a difference in means of 1 unit (specified via the `delta` argument), anticipate a standard deviation of 2 and can afford 10 samples per group.

```
> power.t.test(n = 10, delta = 1, sd = 2)

    Two-sample t test power calculation

              n = 10
          delta = 1
             sd = 2
      sig.level = 0.05
          power = 0.1838375
    alternative = two.sided

NOTE: n is number in *each* group
```

In this scenario, we would have a power of 18.4% to detect a group difference if it really existed. Given this low probability, we would either abandon our research plans or increase the sample size to gain more power. But how many samples are enough? To answer this question, we set the desired power to 80% and rerun the function without specifying the `n` argument.

```
> power.t.test(delta = 1, sd = 2, power = 0.8)

     Two-sample t test power calculation

              n = 63.76576
          delta = 1
             sd = 2
      sig.level = 0.05
```

```
          power = 0.8
    alternative = two.sided

NOTE: n is number in *each* group
```

The output tells us that we need 64 samples per group to raise the power to 80%.

6

Communicating Quantitative Information Using Visuals

Humans are inherently visual beings. There is plenty of research proving the superiority of visuals over text when it comes to the uptaking and retaining information. For example, we process visual information about 60,000 times faster than text. It is far easier to recall visual information than text. In fact, we often translate text to visuals in our memories – this happens if we read a good novel for example. In data-driven scientific disciplines, the phenomenon is exacerbated because the information we convey tends to be rather complex, such that visuals become even more important. However, it is both easy to distort information deliberately and also to inadvertently confuse by using visuals. The former is obviously up to us and relies on the individual researcher's integrity. The latter firstly requires a good understanding of the data underlying the visual representation. Too often, researchers take to plotting without a comprehensive understanding of the data. This is particularly dangerous where software packages tempt us with relatively fancy looking plots that form at a mouse click. Secondly, we need to learn a set of basic graphic rules to make visuals both appealing and scientifically correct. Producing highest-quality, publication-ready graphics is not only beyond the scope of this book but also it occasionally requires us to rely on professionals or at least get their advice.

Before you plot, ensure you know what you want to achieve, often it helps to sketch possible visualisations of your dataset on a piece of paper first

Two fundamental strategies exist when it comes to plotting data, both come with their pros and cons. The first one is to start with a blank slate, adding all elements of a plot by hand (the painter's model). This means the software does not 'tell you what you should do', but rather waits for you to take action, forcing you to reflect on what you actually want to show and what the best figure design is to convey your message. The second one is quite opposite and uses templates, following the philosophy of 'let me suggest something'. Spreadsheet software, for example will readily suggest a plot if you highlight some data and click on the plotting symbols. Whether it is what you want is a different question, and there is also the danger of using a plot without understanding it thoroughly. The R software generally follows the first of those two philosophies, where you add incrementally to a plot, using simple graphic functions, although the *ggplot2* package has somewhat filled the demand for 'template-based' plotting in R, almost creating a parallel world to the traditional R plotting functions. Because there are now so many ways of drawing even the most basic graphs, we introduce you to use not only traditional R graphics but also *ggplot2* options, where we find them more powerful. Our experience shows that sticking exclusively to either (base R or *ggplot2*) is not practical, and there is a lot of value in learning to code both.

Exploratory plots are used to quickly identify patterns, customised graphs are used for reports and publications

In what follows, we distinguish between high-quality plots, i.e. highly condensed and 'ready to present' graphics, and the fundamentally different exploratory plots, which we

R-ticulate: A Beginner's Guide to Data Analysis for Natural Scientists, First Edition.
Martin Bader and Sebastian Leuzinger.

Companion website: www.wiley.com/go/Bader

use to familiarise ourselves with our data, or to look for patterns, outliers, and correlations. In the latter, we need not consider the use of space, nor will we be concerned about the details of how the graphs are presented. Where we do not specifically refer to 'exploratory plots', we normally mean the more elaborate, customised plots. The tools to produce the two types of plots do not differ much, you will largely use the same functions but specify more arguments for highly customised plots. While this chapter focuses on the technical aspect of plotting only, we will come back to graphical representation of data, which are often used alongside the data analysis, throughout the other chapters. Whenever possible, data analysis should culminate in an informative figure.

In terms of the technical (R-) skills, we rely on the basics learnt in Chapters 1 and 2. Comparable to learning a language, you need a solid base vocabulary and a few tips and tricks to get started. Once you reach that level, googling for a chunk of code you might need to add an extra axis to a plot, for example, will be just as easy as looking up the translation of a word you do not know.

6.1 The Fundamentals of Scientific Plotting

Scientific graphing can be viewed as a bit of an art. Unfortunately, sometimes even published scientific graphs do not meet a very high standard. Edward R. Tufte in his seminal book *The Visual Display of Quantitative Information* (Tufte 2018) notes poignantly: 'Graphical excellence is that which gives to the viewer the greatest number of ideas in the shortest time with the least ink in the smallest space'. This should remind us of Occam's razor presented in Chapter 1, which essentially asks for exactly this. We therefore have to consider (i) the ease of access to the key information we want to convey, (ii) the efficient use of space, and (iii) the simplicity or sleekness of a graphic, trying not to overload it with too many elements. Criterion (i) should be viewed as the most important one, where compromises should not be made, (ii) and (iii) will be in competition with each other, and it may take some experience to strike the right balance. Often, information can be condensed in a smart way, and we will see that multi-panel plots and insets can play an important role in this.

Some preliminary considerations we should allow before starting to plot are:

- Is a graph justified in the first place? For example, if we want to report three light levels that were used in an experiment with insects, do we really need to show them in a bar plot? In such a simple case, a sentence mentioning the three light levels will be more parsimonious and just as easy to understand.
- Which variable is the one of interest and needs to be shown prominently? The response variable is usually shown on the y-axis (the vertical axis).
- Are the data aggregated at the right level? For example, if we measure biodiversity in five tidal rock pools that we consider comparable, we want to show the average across all five pools (replicates), as opposed to the five individual pools in five plots.
- Organise the plot such that the eye can easily *compare* between treatments, levels, etc. This means that we should aim to use the same range limits on our axes, if possible.
- Contrary to some older advice you may come across, do add text and visual cues (arrows, shading, etc.) to your plots. Whatever serves the goals set above (facilitates the information uptake by the reader) should be encouraged. For example, legends should be included inside the plotting area if this makes it easier for the reader.

- Consider colour vision deficiency when using colours (e.g. red-green blindness affects a sizeable portion of people and we should either avoid the red-green combination or carefully select shades of red and green that are more distinguishable for individuals suffering from red-green colourblindness).
- Choose a large (but still proportionate) font size for tick and axes labels where possible. Often, what appears to be readable on screen is too small in print. Where there is space, why not increase the font to make things easier for the reader.

The primary goal of a scientific graph is to summarise information and make it as easily accessible for the reader as possible

6.2 Scatter Plots

Scatter plots are the most elementary plots to visualise data. Even if you only have a single variable, you can still plot it against the 'index', i.e. the position of the values in the variable (from 1 to however many values your variable contains). We have used such 'dot charts' earlier (see Chapter 2). Dot charts (use `dotchart`) are useful mainly to get a visual impression of a single variable, especially in terms of extreme values (potential outliers). The function `plot` is very flexible and 'smart', in the sense that it tries its best to visualise whatever data you feed it. For a single variable for example, it will produce something that looks like a dot chart, plotting the values against their index. For two continuous variables, it will produce a scatter plot, but as soon as one variable is categorical, it will show a box plot (see below).

Try the function `plot` on any object (e.g. data frames, model outputs) – it will produce the most appropriate plot (e.g. pairs plots, diagnostic plots)

Let us put this into practice. To save space we do not print the resulting plots, you are encouraged to use the following code to produce the corresponding plots on your own:

```
> plot(runif(100, 0, 100)) # true random 'night sky', one uniform variable
> dotchart(runif(100, 0, 100)) # the same using dotchart()
> plot(runif(100, 0, 100), runif(100, 0, 100)) # random night sky again,
    two uniform variables
```

In scatter plots, we normally have the predictor variable on the x-axis, and the response variable on the y-axis. If we do not have a designated response variable, the assignment of axes to the two variables will not matter. By default, R uses generic plotting parameters that can accommodate a large variety of variables. For example, this is why the tick labels on the y-axis are vertical. Figure 6.1 illustrates how you could modify a simple scatter plot to make it more visually appealing, for example to get it ready for a report or a publication. We use the in-built dataset `cars` in our example, which contains the trivial relationship between the variable 'speed of a car' and 'distance it takes for it to stop'. The code for the customised plot on the right of Figure 6.1 is:

```
> plot(dist ~ speed, data = cars,
    xlab = "Speed (mph)", ylab = "Distance (ft)", # custom axis labels
    xlim = c(0, 26), ylim = c(0, 122), # custom x and y axis limits
    xaxs = "i", yaxs = "i", # axes set to 'identity', i.e. no offset
    mgp = c(1.2, .2, 0), # defines the distances between the axes, tick
          label, and axis labels
    cex = 0.7,  cex.axis = 0.8, cex.lab = 1, # play with character and font
          sizes
    tcl = 0.3, las = 1, pch = 16) # change tick lengths and choose the
          symbol character
```

If you would like to add data to an existing plot, you can use `points`. The software R makes a distinction between 'high-level plotting functions' (those that produce a *new* plot), and

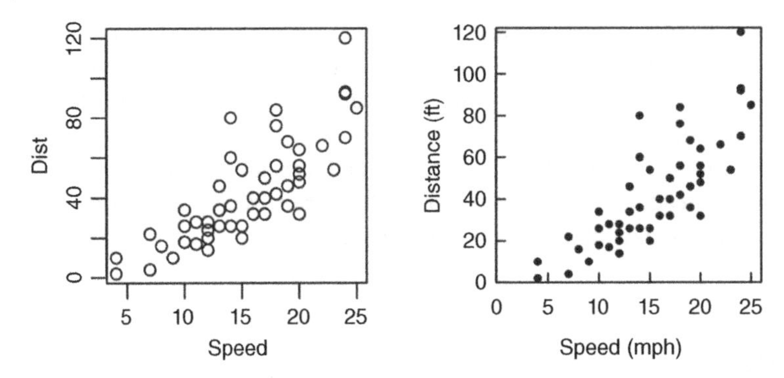

Figure 6.1 Two times the same plot, on the left the way R produces it in its default mode, using the code `plot(dist ~ speed, data = cars)`. On the right, font sizes, tick lengths, tick labels, axes labels, and the symbols have been adjusted to make an aesthetically more appealing plot.

'low-level' plotting functions (those that *add* to an existing plot). Box 6.1 gives an overview of the most common high- and low-level plotting functions.

When using `points`, make sure your x and y values are inside the plotting area. For example if you would like to add the data point of an additional car (30 mph, 125 ft) to the plot in Figure 6.1, you will not be able to see it, simply because it is outside the already set limits of the x- and y-axes (so you would have to adjust the axis limits first to visualise this new data point).

Box 6.1 Plotting functions in R

High-level plotting functions in R initiate a new plot, while low-level plotting functions will add to an existing plot. If used without an existing plot, low-level plotting functions will logically evoke an error message telling you that the parent plot does not (yet) exist. Below is a summary of a set of functions that will often be used both in exploratory plotting, as well as for highly customised plots.

High-level plotting functions

`plot`	Very generic, will produce the most suitable plot given the argument you provide
`barplot`	Very flexible function to produce all sorts of horizontal or vertical bar plots
`hist`	Produces a histogram
`boxplot`	Mostly used for exploratory plotting when you have a categorical predictor and a continuous response, computes the median and lower and upper quartiles
`pairs`	Exploratory plotting function to show all relationships between all variables
`qqnorm`	Plots the quantiles of a standard normal distribution against those of the variable provided, used to assess the normality of a variable
`image`	Used for heat maps

Low-level plotting functions

`points`	Adds data points to an existing plot
`lines`	Adds data points joined by a line to existing plots

`rect`	Adds rectangles to an existing plot
`arrows`	Adds arrows (or error bars) to an existing plot
`segments`	Adds straight lines to an existing plot
`abline`	Adds a straight line to a plot, often used to add horizontal or vertical lines, or to add a model fit of a linear regression
`polygon`	Adds any polygon to an existing plot, can for example be used to add shaded error bands to a fitted line
`axis`	Adds an axis to an existing plot (at the bottom, left, top, or right)
`text` and `mtext`	Adds text to a plot (`text`) or to the margin of a plot (`mtext`), often used to add axis titles or information on top of a figure
`legend`	Adds a legend to an existing plot (within the plotting area)
`title`	Adds a title to the plot
`box`	Draws a box around an existing plot (often used when axes are suppressed in the original plot)

If we have several (mostly continuous) variables in a dataset, the `pairs` plot is extremely useful, as it shows, at a glance, the relationship between all combinations of variables. Let us take the dataset 'swiss' which comes with R. Try

```
> pairs(swiss)
```

to see an example. You will see that `plot` in fact produces the same plot (remember, `plot` intends to produce the most sensible plot given the argument you provide). There are plenty of opportunities to pack more information into a pairs plot, see the `pairs` help file by typing `?pairs`. The *ggplot2* (Wickham 2016) alternative is worth having a look at:

```
> install.packages("ggplot2") # packages need to be installed only once
> install.packages("GGally")
> library(ggplot2)# load ggplot2 package
> library(GGally) # load GGally package
> ggpairs(swiss)
```

It additionally provides the distributions of the variables along the diagonal, and the correlation coefficients in the upper part of the panel. More details on 'pairs' plots will be provided in Chapter 8 on correlation analysis.

6.3 Line Plots

If you 'join the dots' on a scatter plot, you will obtain a line graph. Where and whether this makes sense has to be considered carefully. We have to be cautious not to imply that we have observations in between the data points that we join. A good way to show this is to not connect the dots to the lines in between, as R does by default. As an example, let us use the dataset `Orange` that comes with R. Because it contains diameter growth of five trees and their ages, it is reasonable to assume that in between the available data points, the trees will not shrink, but rather show a more or less linear growth pattern. A quick exploratory plot to show the growth of tree 1 in this dataset could be produced using

```
> plot(circumference ~ age, data = Orange[Orange$Tree == 1, ], type = "l")
```

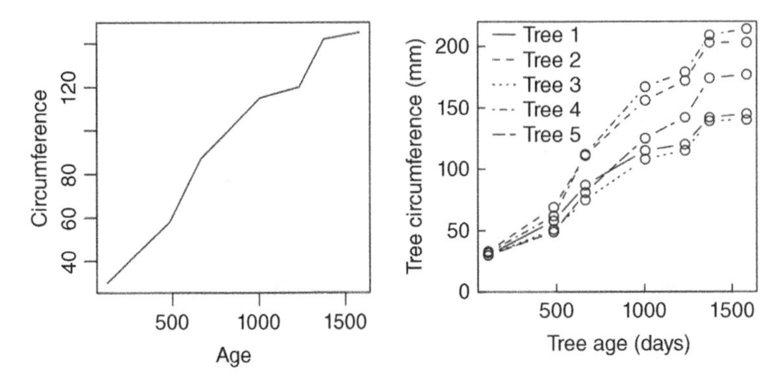

Figure 6.2 An exploratory, quick line plot (left), and a more elaborate plot using the in-built dataset Orange, which contains the age and diameter of five orange trees.

The function plot also accepts age and circumference to be supplied as *x* and *y* values, but this makes indexing cumbersome, as the data argument only works if you use the formula format (using the ~ symbol, which is called 'tilde' and reads 'as a function of'):

```
> plot(x = Orange$age[Orange$Tree == 1], y = Orange$circumference
        [Orange$Tree == 1], type = "l")
```

Tree number 1 is selected using the square brackets and logical matching (read this as 'choose the rows where the variable "Tree" within the dataset "Orange" equals 1', and note the comma in the square brackets where the whole dataset is selected, see Chapter 2). The only other argument we have to set is 't' (the type of plot, 'l' here stands for 'line', 'p' for points, and 'b' for both lines and points). The resulting plot is shown on the left hand side of Figure 6.2. If we want to construct a more elaborate plot, for example with the age–diameter relationship of all five trees on one graph including a legend, we may use the below code. The argument 'lty' stands for 'line type' and lets us choose different types of lines. Most of the other arguments used are similar to those used earlier in the scatter plot (Figure 6.1). The resulting plot is shown on the right side of Figure 6.2.

To avoid implying that we have data where we do not, leave a gap between the data point and the connecting line

```
> par(mfrow = c(1, 2))
> plot(circumference ~ age, data = Orange[Orange$Tree == 1, ], t = "l")
> plot(circumference ~ age, data = Orange[Orange$Tree == 1, ], type = "b",
       ylim = c(0, 220), xlab = "Tree age (days)", ylab = "Tree
       diameter (mm)", yaxs = "i", mgp = c(1.3, .2, 0), cex.axis = 0.8,
       cex.lab = 1, tcl = 0.3, cex = 0.7,  las = 1)
> lines(circumference ~ age, data = Orange[Orange$Tree == 2], type = "b",
        lty = 2)
> lines(circumference ~ age, data = Orange[Orange$Tree == 3], type = "b",
        lty = 3)
> lines(circumference ~ age, data = Orange[Orange$Tree == 4], type = "b",
        lty = 4)
> lines(circumference ~ age, data = Orange[Orange$Tree == 5], type = "b",
        lty = 5)
> legend(0, 220, legend = c('Tree 1', 'Tree 2', 'Tree 3', 'Tree 4',
         'Tree 5'), lty = 1:5, bty = "n")
```

The four lines commands can nicely be packed into a for loop for greater efficiency, see Box 10.1 for the use of loops:

```
> for (i in 2:5) {
> lines(circumference ~ age, data = Orange[Orange$Tree == i], type = "b",
        lty = i)}
```

6.4 Box Plots and Bar Plots

Box plots and bar plots are related but used for slightly different purposes. Both usually require a categorical predictor variable and a continuous response variable to make sense. Box plots let us compare both the distribution within, and the difference between, groups. They are mostly used to quickly explore datasets with a continuous response variable and at least one grouping variable (a factor or categorical predictor). To illustrate how a box plot works, we again use an in-built dataset `iris`. Try for example:

```
> boxplot(Sepal.Length ~ Species, data = iris)
```

This will show you how the variable 'sepal length' varies between species, but it also gives you an idea of the distribution of the variable within the groups. Note that this type of plot shows you the median and the first and third quartiles for each group (here species) as the center and the upper and lower edges of the box. This allows you to see that 'sepal length' is not much skewed in all three species and seems to vary relatively symmetrically around the median. The *ggplot2* package offers a variety of options to produce boxplots, one that is complementary and worth mentioning is the so-called 'violin' plot, which shows the distribution of the data within a group using a probability density fit (Figure 6.3). So you can think of a violin plot as an upended and mirrored density histogram. Often, violin plots are combined with boxplots:

```
## Violin plot using ggplot2
> library(ggplot2)
> ggplot(iris, aes(x = Species, y = Sepal.Length)) +
  geom_violin() + geom_boxplot(width = 0.2)
```

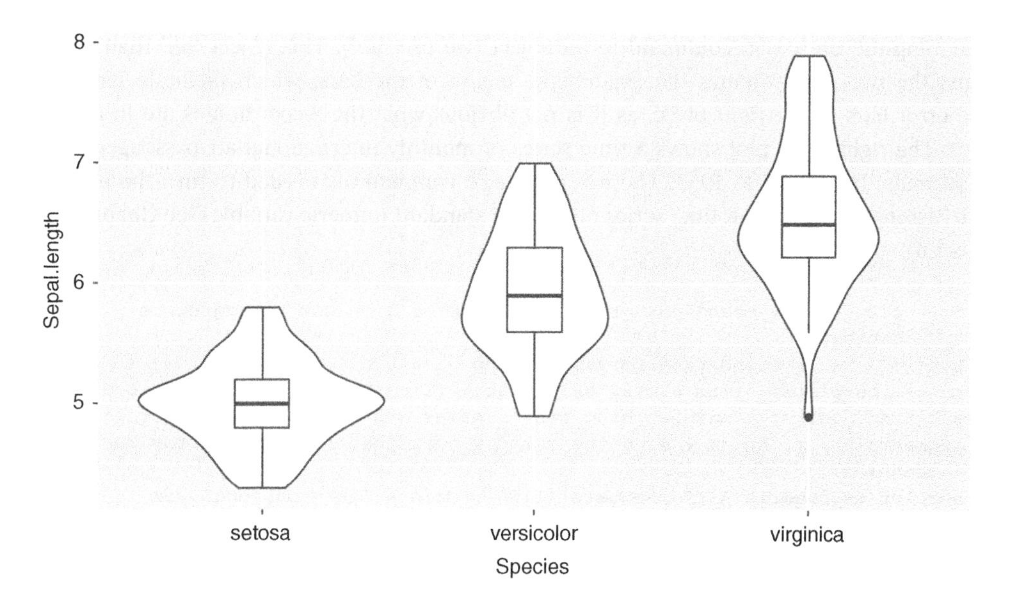

Figure 6.3 A simple version of a box and violin plot using the *ggplot2* package on the built in `iris` dataset, visualising the differences in sepal length by species. The 'violins' also provide insight into the distribution of the values within species. For example, they show an increase in frequency of high values around 7.7 in the species 'virginica', something that cannot be seen from the boxplots alone.

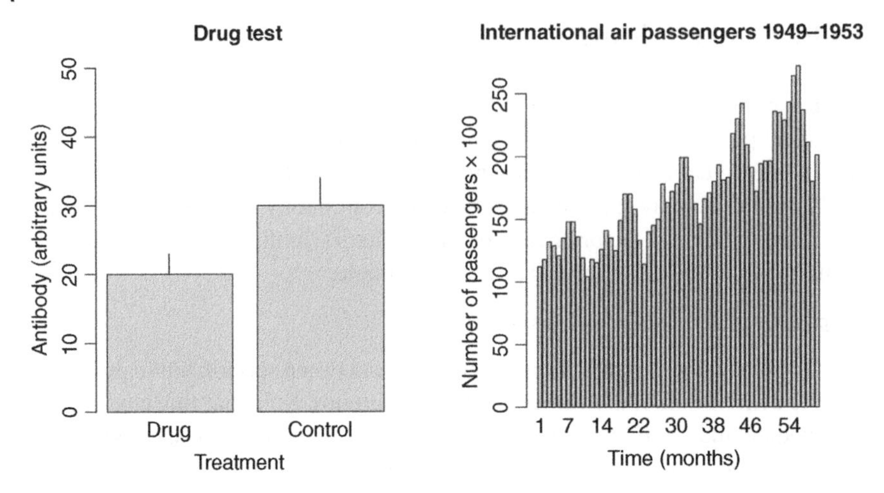

Figure 6.4 Different applications of simple bar plots. Left: A comparison between two groups using the mean and standard error. Right: A 60 months time series of international air passengers from an in-built dataset (AirPassengers).

Bar plots are different in that they usually show the group means only, often with the standard errors (or 95 % confidence bounds) added as 'antennas' on top of the bars. They can also be useful to show quantities over time, for example daily accumulated rain over a month or a year.

Two examples are shown in Figure 6.4. On the left, we show the simplest form of a bar plot that is used to compare two groups (we use a drug test, where the first group receives the actual drug, and the second one acts as a control). Note the little trick of assigning the barplot command to an object (we use 'hh'). This object 'hh' then contains the two *x*-coordinates that match the centre of the bars, which facilitate placing the error bars in the right place, as it is not obvious what the *x*-coordinates are in a bar plot. The right hand plot shows a time series of monthly international air passengers (in thousands) from 1949 to 1953. The as.numeric command is needed to turn the object 'AirPassengers' from an R time series object to a standard numeric variable (see Chapter 1, Box 1.4).

```
> m <- c(20, 30) # means per group, see chapter 2 on how to aggregate
       datasets
> s <- c(3, 4) # standard errors per group
> hh <-  barplot(m, ylim = c(0, 50), xlab = 'Treatment', ylab = 'Antibody
  (arbitrary units)', main = 'Drug test', names.arg = c('Drug', 'Control'))
> segments(hh, m, hh, m + s) # the x and y coordinates for the error bars
  ('antennas')
> barplot(as.numeric(AirPassengers)[1:60], main = 'International air
          passengers 1949-1953', ylab = 'Number of passengers \u00D7 1000',
          xlab = 'Time (months)', names.arg = 1:60) # for \u00D7 see Box 6.1
```

Box plots are often used to explore a dataset with a categorical predictor, bar plots serve to compare means and to visualise time series data

The function barplot also allows you to vary the widths of the bars, the colours, and you can stack bars on top of each other (for example to show percentages out of 100). Type 'example("barplot")' to get an impression.

6.5 Multipanel Plots and Plotting Regions

Multiple equally sized graph panels can be arranged on a single page using the `mfrow` argument of the `par` function. The `par` command stands short for parameter settings, and can be used to set a large number of graphical parameters. The argument `mfrow` asks for a numeric object of the form '`c(number of rows, number of columns)`', which determines the number of panels on a page. Try these lines of code:

```
> x <- rnorm(n = 50)
> y <- rnorm(n = 50)
> lm1 <- lm(y ~ x)
> par(mfrow = c(2, 2)) # four plots in two rows and two columns
> plot(lm1)
> par(mfrow = c(1, 4)) # four plots in one row
> plot(lm1) # diagnostic plots for lm1 (see Chapter 9)
```

The more flexible option to create highly customised multi-panel plots and insets is the function `layout`, as it allows the arrangement of differently sized panels within an overarching figure. The underlying idea is that you divide the plotting area into a number of rows and columns, where all row heights and column widths can be controlled individually, and any graph can occupy more than one row or more than one column. Figure 6.5 shows an example in which an inset graph is produced using the following code:

```
> layout(mat = matrix(c(1, 1, 1,
                        1, 2, 1,
                        1, 1, 1), nrow = 3),
       widths = c(2, 1.3, 0.2),
       heights = c(1, 1.5, 1))
> layout.show(2) # visualises the layout (intended for format checking only,
    so disable this command when you put together a real multi-panel
    figure)
> x <- rchisq(n = 1000, df = 1)
> hist(x, main = "")
> plot(x)
```

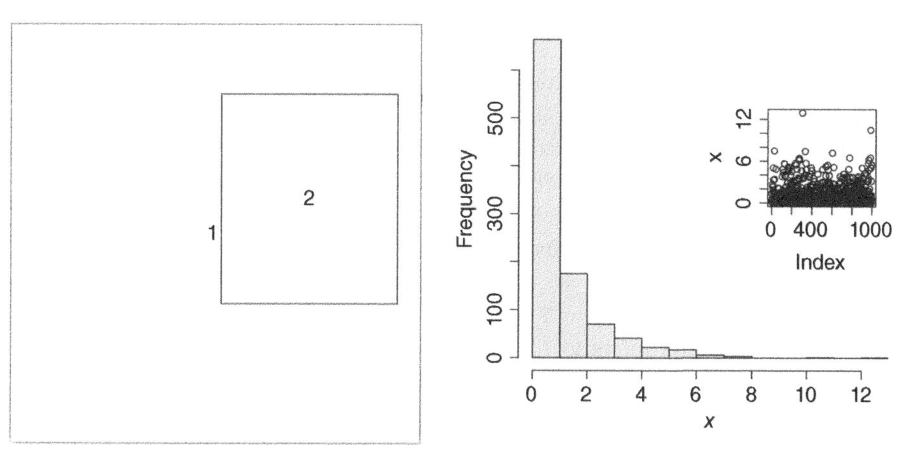

Figure 6.5 The layout command lets us divide the plotting area into as many subplots as desired. The left side of the figure shows the output of `layout.show(2)`, which is used to visualise the two plotting areas.

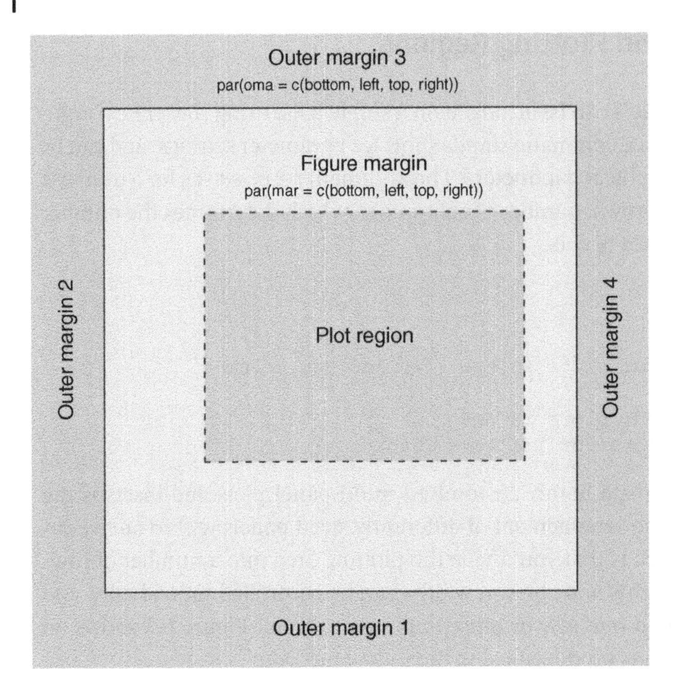

Figure 6.6 Inner and outer margins in R figures. It is important to understand how margins are organised, for example, if you want to place text into the margins.

It is extremely useful to know about figure margins, for example if you would like to print text into the margins or reduce them to minimise white space between panels. Figure 6.6 gives an overview of how margins are organised, and how to change them using the 'mar' and 'oma' arguments of the function `par`.

6.6 Adding Text, Formulae, and Colour

Of course, R lets you do almost everything in terms of customising your plots with special characters, formulae, colour, etc. A bit like with a box of Lego, you can build your objects or graphs using different tools, for example by using the standard R functions, or the *ggplot2* suite of functions. For adding text, there are two main functions: `text` is used for annotations in the plotting area while `mtext` (which stands for margin text) may be used for annotations in the figure margin (e.g. custom axis labels, additional text above a graph). Use '`mtext("your text...", side = ...)`' for the figure margin, and '`mtext("your text...", side = ..., outer = T)`' to add text into the outer margin (see Figure 6.5). The argument `side` always refers to 1 = bottom, 2 = left, 3 = top, and 4 = right. The `text` function also includes arguments to adjust (`adj`) and rotate (`srt`) text. The tricky bit starts when subscripts, superscripts, multi-line text, and special characters are required. Some of these are shown not only in Box 6.2, but also type `?plotmath` in your console to call up a comprehensive list of special characters in the help file. Online resources are plentiful.

Box 6.2 Special characters and fonts in R. You will see that virtually everything can be achieved, and often, searching the internet gets you there quickest. Use the following code to learn a series of functions and tricks to add text to a plot. Note how we plot 'nothing' (specified as NA), in order to get a blank canvas, which we then use to add elements of text and formulae to. This strategy lends itself for highly customised plots, where you want to have full control over every aspect of your plot. The function `box` is used to draw a box around the plotting area. We suggest that you use the provided code, then change one argument at a time and watch the resulting change. The resulting plot is shown.

```
> plot(NA, xlim = c(0, 100), ylim = c(0, 10), ann = F, axes = F) # blank plot
> box()
> mtext("Auckland, New Zealand", side = 3, line = 0.5, at = 0, adj = 0)
> text(x = 20, y = 9.5, "Learning R syntax isn't hard,\n and it will pay off!")
> ## Use x and y to specify position and "\n" to insert a line break
> par(lheight = 2) # double line spacing
> text(x = 20, y = 8, "Learning R syntax isn't hard,\n and it will pay off!")
> par(lheight = 1) # back to single spacing
> text(20, 6, "UseR!", srt = 45) # srt: string rotation in degrees
> text(20, 3.5, "UseR!", srt = 90)
> text(20, 1, "UseR!", srt = -45)
> text(70, 9.5, expression(paste("r"^2, " = 0.87"))) # superscript
> text(70, 8.5, expression("CO"[2])) # subscript
> text(70, 7.5, expression(hat(y) %+-% se)) # hat and plus/minus
> text(70, 6, expression(bar(x) == sum(frac(x[i], n), i == 1, n)))
> text(70, 3, expression(bar(x) == sum(frac(x[i], n), i == 1, n)),  cex = 3)
> text(70, 6, expression(bar(x) == sum(frac(x[i], n), i == 1, n)))
> text(70, 3, expression(bar(x) == sum(frac(x[i], n), i == 1, n)), cex = 3)
> text(95, 5, 'Number of passengers \u00D7 1000', srt = 90)
```

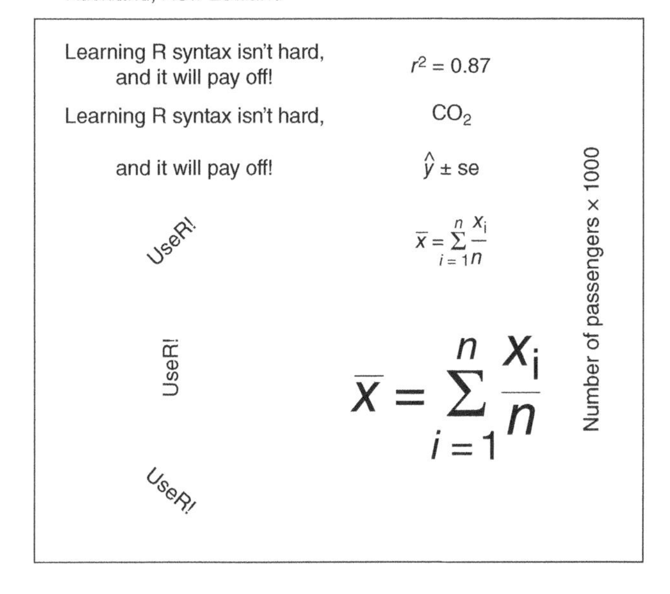

Legends are drawn with the function `legend`, or you can assemble them using `segments`, `rect`, `points`, and `text`, etc. In most cases, using the function `legend` will be the quicker way to go:

```
> x <- 1:50
> y <- rnorm(n = 50, mean = 10)
> plot(y ~ x, xlim = c(0, 50), ylim = c(0, 20), pch = 1) # an example plot
> points(x, rnorm(50, 5), pch = 16) # using different point characters
    ('pch')
> treat <- expression(paste("NO"[3],""^"-")) # legend entry with sub- and
    superscript
> legend(x = 0, y = 20, c("Control", treat), pch = c(1,16))
> legend(x = 15, y = 20, c("Control", treat), pch = c(1,16), bty = "n")
    # no box
```

6.7 Interaction Plots

Statistical interactions i.e. relationships between two variables that are dependent on a third predictor variable are both common and easily misunderstood or overlooked. Because verbalising interactions can be very difficult, having good visual tools available to illustrate such statistical results is extremely important. The interpretation of interactions is treated in Chapters 8, 10, and 11, so here we will purely focus on the technicalities of how to produce the plots. A very simple agricultural example in which we show that fertiliser application depends on irrigation shall be used:

```
> set.seed(9) # a made up dataset
> d1 <- data.frame(yield = rnorm(40, mean = 20),
           fert = factor(rep(c('N', 'ctrl'), 20)),
           water = factor(rep(c('water', 'ctrl'), each = 20))
           )
```

To quickly show how the response of yield to fertiliser addition depends on irrigation, we can use `coplot` from base R. The function is designed to provide a quick exploratory plot rather than a customised plot you would show in a report (left panels of Figure 6.7). The middle and right panels of Figure 6.7 are two variants of an interaction plot produced by `ggplot`. The code for all three panels is as follows:

```
## Base R coplot
> coplot(yield ~ fert | water, data = d1)
## The equivalent ggplot
> ggplot(data = d1, aes(x = fert, y = yield)) + geom_point() + facet_wrap
    (facets = ~ water)
## Similar but with a boxplot
> ggplot(data = d1, aes(x = fert, y = yield)) +  geom_point() +
     geom_boxplot() + facet_wrap(facets = ~ water)
```

6.8 Images, Colour Contour Plots, and 3D Plots

6.8.1 Adding Images to Plots

In scientific plotting, we should be cautious adding images, 3D elements, and other visual effects. The main goal must always be to make the information as easily accessible for the reader as possible, without distracting from the main message, and without distorting

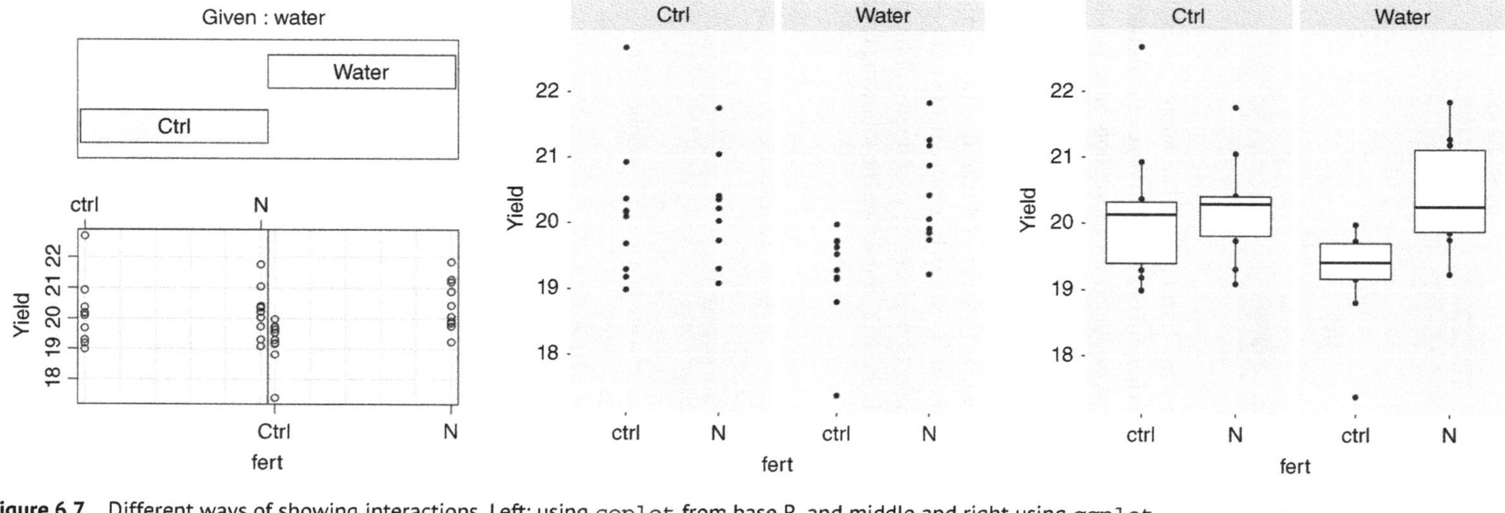

Figure 6.7 Different ways of showing interactions. Left: using `coplot` from base R, and middle and right using `ggplot`.

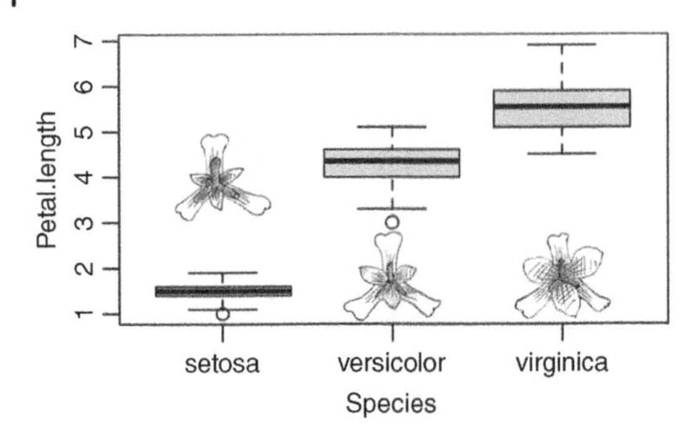

Figure 6.8 Inserting pictures into a plot using the functions `readPNG` and `rasterImage`.

the facts. We should always follow the 'Principle of Parsimony' as explained in detail in Chapter 1. Sometimes however, it is helpful to add graphical elements that go beyond simple, two-dimensional line, bar, or dot plots. Reach for those tools once you have exhausted the palette of traditional plotting tools.

Inserting a picture or symbol into a figure can be useful to directly illustrate subjects or items a plot is directly associated with. Probably the easiest way to do this is using the package *png*. Place any .png file into your work directory and read it with `readPNG`. The picture is then stored in your object (e.g. '`setosa`') and can be plotted using `rasterImage`. To place your picture where you want it, refer to the existing coordinates of the plot. Here, the mid points of the boxes along the x-axis are 1, 2, and 3. See the following code and the resulting plot in Figure 6.8.

```
> library(png)
> setosa <- readPNG('setosa.png') # read in png pictures
> versicolor <- readPNG('versicolor.png')
> virginica <- readPNG('virginica.png')
> boxplot(Petal.Length ~ Species, data = iris)
> ## Place the picture using plot coordinates
> rasterImage(setosa, xleft = .7, ybottom = 3, xright = 1.3, ytop = 5)
> rasterImage(versicolor, 1.7, .8, 2.3, 2.8)
> rasterImage(virginica, 2.7, .8, 3.3, 2.8)
```

6.8.2 Colour Contour Plots

Contour plots (with or without colour gradient) are a great tool for displaying geographical contours and also for visualising 3D data in a 2D graph.

For briefly illustrating how you can create topographic maps featuring contour lines, we load the in-built dataset `volcano`, which is simply a matrix of size 87×61.

```
> data(volcano)
> class(volcano); dim(volcano)
[1] "matrix"
[1] 87 61
```

To visualise this matrix (which by the way represents the dormant volcano Maungawhau in New Zealand), we can use `image`, `contour` or both (Figure 6.9).

```
> contour(volcano) # contours only, see Figure 6.9 left
> image(volcano) # coloured surface
> image(volcano); contour(volcano, add = T) # both, see
    Figure 6.9 right
```

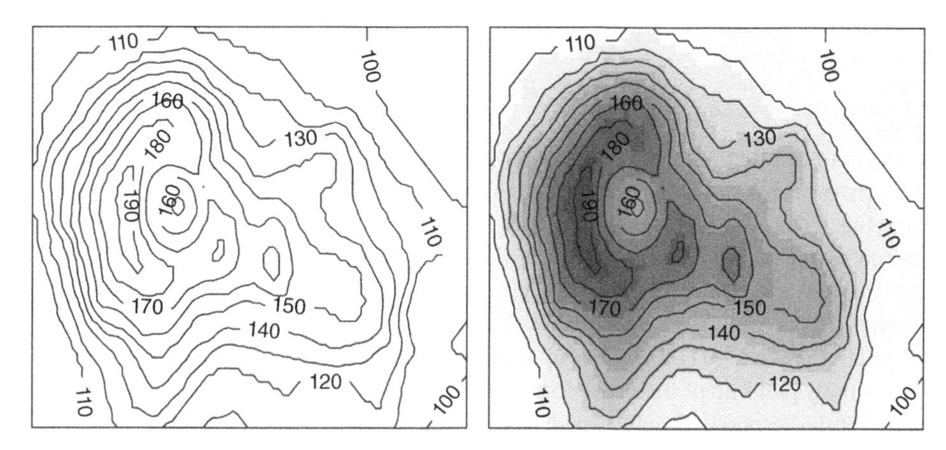

Figure 6.9 Contour plot (left) and colour contour plot (right) of the dormant volcano Maungawhau, Auckland, New Zealand (built-in `volcano` dataset).

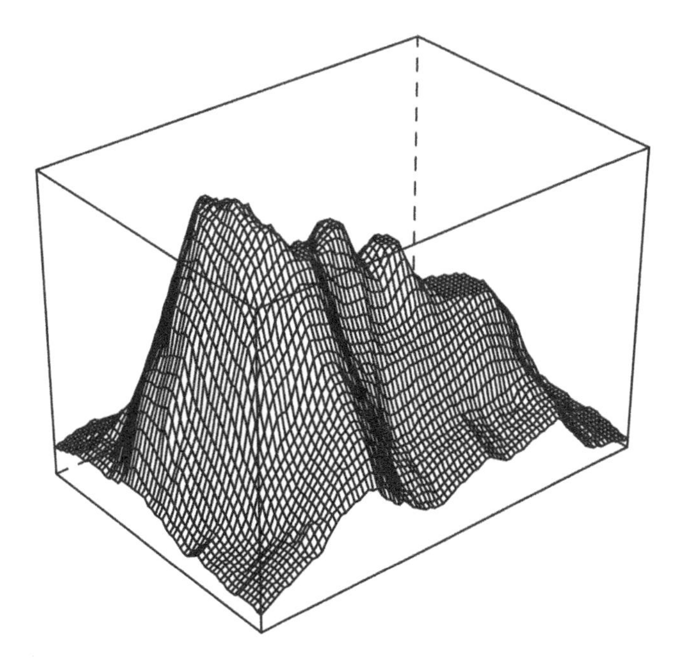

Figure 6.10 3D plot of Mt. Eden (Maungawhau), Auckland, New Zealand (built-in `volcano` dataset) using the `cloud` and `wireframe` functions in R package *lattice*.

After loading the package *lattice* (Sarkar 2008), the functions `cloud` and `wireframe` can be used to create a 3D visualisation (Figure 6.10).

```
> library(lattice) # loading the lattice library

> cloud(volcano) # 3D plot
> wireframe(volcano) # wireframe plot
```

If you want to change the angle at which you are looking at your wireframe, you can use the `persp` function:

```
> x <- 1:nrow(volcano)
> y <- 1:ncol(volcano)
> persp(x, y, volcano, theta = 110, phi = 40) # you can play
        with theta and phi
```

Drawing maps is often required but goes beyond the scope of this book, and we refer the reader to the numerous online resources or the excellent book *Geocomputation with R* by Lovelace et al. (2019).

Colour contour plots are also a great tool for visualising relationships between a continuous response variable and two continuous predictors. In the latter case, the contours represent isolines that connect points of equal value. This means that colour contour plots provide an elegant alternative to 3D plots. Instead of plotting the response variable on the *y*-axis as usual, we simply plot one of the predictors on the *y*-axis and the second one on the *x*-axis. The values of the response variable are displayed as a colour gradient. However, using the raw data rarely works since the plotting functions require a single response value (*z* variable) at each combination of *x* and *y*. Also, the latter should ideally be regularly spaced along their ranges, which is often not the case. The best results are commonly achieved by providing an interpolated spline surface of our raw data (R package *MBA*; Finley et al. 2022) or a set of model predictions to create this type plot. We exemplify the creation of a contour plot using a time series of a summertime vertical water temperature profile of one of the lakes of the famous IISD Experimental Lakes Area (International Institute for Sustainable Development), a unique research environment in North-western Ontario, Canada (Pilla et al. 2021) (Figure 6.11). First, we create a spline-based surface approximation of our data

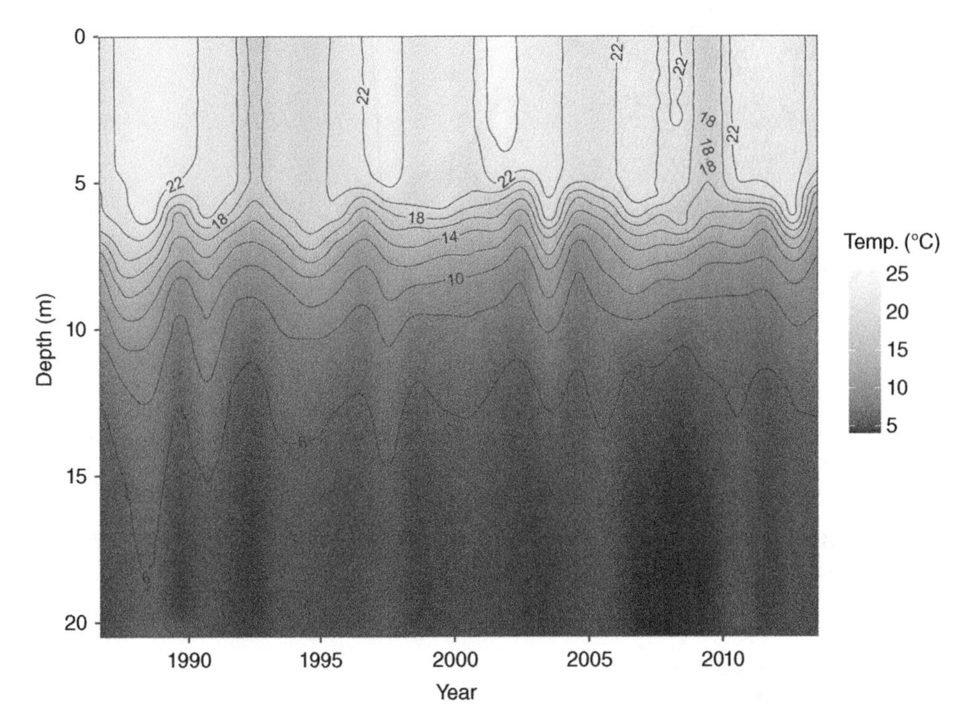

Figure 6.11 Colour contour plot of water temperature as a function of depth and sampling year in one of the lakes of the famous IISD Experimental Lakes Area (International Institute for Sustainable Development), Ontario, Canada. Note how the two predictor variables *water depth* and *sampling year* are drawn on the *y*- and *x*-axes, while the response variable *water temperature* is displayed as a greyscale colour palette. Please note that the E-book version contains a rainbow colour palette (turbo palette in R package *viridis*).

using the `mba.surf` function with a resolution of 500 equally spaced values along the *x*- and *y*-axes to ensure a smooth image. The resulting 500×500 matrix yields a data frame with 250,000 rows. We then plot a colour gradient of water temperature (response variable) as a function of measurement date (*x*-axis) and water depth (*y*-axis) as shown in Figure 6.11. The workhorse function in the traditional graphics system for this type of plot is called `image` and can directly use the matrix returned by the `mba.surf` function. A quick conversion from matrix to a data frame in long format is required for plotting in the *ggplot2* framework. We outline these steps for both graphics systems below. We will need the packages ***metR*** (Campitelli 2021) and *reshape2* (Wickham 2007).

```
## Load required packages
> library(ggplot2)
> library(MBA) # contains the mba.surf function for B-spline
      interpolation (surface approximation)
> library(metR) # contains improved contour lines function for
      gpplot2
> library(reshape2) # contains the melt function needed to
      convert a matrix into long data frame format
> library(viridisLite) # contains great colour palettes

## Read in lake temperature profile time series
> el <- read.csv("experimental_lake.csv") # el =
      experimental lake
> str(el)
'data.frame':      1161 obs. of  6 variables:
 $ siteid  : int  18 18 18 18 18 18 18 18 18 18 ...
 $ lakeid  : int  61 61 61 61 61 61 61 61 61 61 ...
 $ lakename: chr  "IISD Experimental Lake 373" "IISD
              Experimental Lake 373" ...
 $ date    : chr  "1986-08-05" "1986-08-05" "1986-08-05"
              "1986-08-05" ...
 $ depth   : num  0 0.5 1 1.5 2 2.5 3 3.5 4 4.5 ...
 $ temp    : num  20.2 20.2 20.2 20.2 20.2 ...
```

In the structure output, the date shows up as a character string, which should rather be formatted as a proper date. Rerunning the `str` command after the recoding step below, will show the change.

```
> el$date <- as.Date(el$date) # apply proper date format
> head(el, n = 3)
  siteid lakeid                   lakename       date depth temp
1     18     61 IISD Experimental Lake 373 1986-08-05   0.0 20.2
2     18     61 IISD Experimental Lake 373 1986-08-05   0.5 20.2
3     18     61 IISD Experimental Lake 373 1986-08-05   1.0 20.2

## The input data frame for the mba.surf function should only contain the
   three relevant variables
> dat <- el[ , c("date", "depth", "temp")]

## Dates need to be expressed numerically to ensure smooth integration with
   the mba.surf function
> dat$date <- as.numeric(dat$date)

## Assign new variable names to comply with the xyz matrix notation of the
   three dimensions
> names(dat) <- c("x", "y", "z")

## Surface approximation using multilevel B-splines ('multilevel' means a
```

```
  coarse to fine hierarchy is applied)
> dat.surf <- mba.surf(xyz = dat, no.X = 500, no.Y = 500, extend = T)

## Add the x- and y-values (date, depth) as dimension names to the matrix of
   z-values (water temperature)
> dimnames(dat.surf$xyz.est$z) <- list(dat.surf$xyz.est$x,
      dat.surf$xyz.est$y)

## Colour contour plot using traditional graphics
## Pre-define axis ranges
> xlim <- range(dat$x)
> ylim <- rev(range(dat$y)) # reverse scale to start with 0 m depth at
      the top

## Adjust the graphical parameters related to the plot and outer margins for
   multi-panel plotting (remove margins around the individuals plot and
   add outer margins to the multi-panel graph)
op <- par(mar = rep(0, 4), oma = c(4, 4, 1, 3))

## Multi-panel layout for two plots (main plot, colour bar legend)
> layout(rbind(c(1, 0, 0),
               c(1, 0, 2),
               c(1, 0, 0)), widths = c(0.95, lcm(0.5), 0.05))

## Main plot
> image(x = dat.surf$xyz.est$x, y = dat.surf$xyz.est$y,
      z = dat.surf$xyz.est$z, col = turbo(n = 200), ylim = ylim,
      yaxs = "i", xlim = xlim, xaxs = "i", zlim = range(dat.surf$xyz.est$z,
      na.rm = T), ann = F, axes = F)
> contour(x = dat.surf$xyz.est$x, y = dat.surf$xyz.est$y,
      z = dat.surf$xyz.est$z, col = "grey15", ylim = ylim,
      yaxs = "i", xlim = xlim, xaxs = "i", ann = F, add = T,
      lwd = 0.5, drawlabels = T, lty = 1, labcex = 0.7)
> axis(2, las = 1, mgp = c(3, 0.6, 0), tcl = -0.3)
> axis.Date(side = 1, x = dat$date, mgp = c(3, 0.5, 0),
      tcl = -0.3)
> box()
> mtext("Depth (m)", side = 2, line = 2, cex = 0.9)
> mtext("Year", side = 1, line = 2, cex = 0.9)

## Colour bar (legend)
> image(x = 4, y = seq(min(dat.surf$xyz.est$z), max(dat.surf$xyz.est$z),
      length.out = 200), z = matrix(1:200, nrow = 1), col = turbo(n = 200),
      axes = F, ylim = c(min(dat.surf$xyz.est$z), max(dat.surf$xyz.est$z)),
      yaxs = "i", ann = F)
> axis(side = 2, at = seq(5, 25, by = 5), labels = F, tcl = 0.4)
> axis(side = 4, at = seq(0, 25, by = 5), labels = seq(0, 25, by = 5),
      las = 1, tcl = 0.4, mgp = c(3, 0.4, 0))
> box()
> mtext(expression(paste("Temp. (", degree, "C)")), side = 3, line = 0.4,
      cex = 0.75, adj = 0)
> par(op)


## Colour contour plot using ggplot2

## Convert the matrix output of the mba.surf into a data frame (long format)
   for plotting with ggplot
```

```
> dat.mba <- melt(dat.surf$xyz.est$z, varnames = c("date", "depth"),
      value.name = "temp")

> head(dat.mba)
      date depth     temp
1 6060.000     0 20.20000
2 6079.737     0 20.30153
3 6099.475     0 20.42318

## Convert date back into a proper date format for plotting
> dat.mba$year <- as.Date(dat.mba$year)

> ggplot(data = dat.mba, aes(x = date, y = depth, z = temp)) +
  geom_raster(aes(fill = temp), interpolate = T) +
  # scale_fill_gradientn(name = expression(paste("Temp. (", degree, "C)")),
      colours = grey.colors(n = 200, start = 0.1, end = 0.95)) +
  scale_fill_viridis_c(name = expression(paste("Temp. (", degree, "C)")),
      option = "turbo") + # the _c denotes a continuous scale
  geom_contour2(aes(z = temp, label = after_stat(level)), size = 0.2,
      colour = "grey15", label_size = 3) +
  scale_x_date(name = "Year", limits = c(min(dat.mba$date),
      max(dat.mba$date)), expand = c(0, 0)) +
  scale_y_reverse(name = "Depth (m)", breaks = seq(0, 20, by = 5),
      limits = c(max(dat.mba$depth), 0), expand = c(0, 0)) +
  theme_bw()
```

References

Campitelli, E. (2021) Tools for Easier Analysis of Meteorological Fields. R package version 0.14.0. https://github.com/eliocamp/metR.

Finley, A, Banerjee, S, and Hjelle, Ø (2022). MBA: Multilevel B-Spline Approximation. R package version 0.1-0. https://CRAN.R-project.org/package=MBA.

Garnier, S., Ross, N., Rudis, R. et al. (2023). viridis(Lite) - Colorblind-Friendly Color Maps for R. viridisLite package version 0.4.2. https://sjmgarnier.github.io/viridis/

Lovelace, R., Nowosad, J., and Muenchow, J. (2019). *Geocomputation with R.* CRC Press.

Pilla, R.M., Mette, E.M., Williamson, C.E. et al. (2021). Global dataset of long-term summertime vertical temperature profiles in 153 lakes. *Scientific Data* 8: 200. https://doi.org/10.1038/s41597-021-00983-y.

Sarkar, D. (2008). *Lattice: Multivariate Data Visualization with R.* New York: Springer. ISBN 978-0-387-75968-5, http://lmdvr.r-forge.r-project.org.

Tufte, E.R. (2018). *The Visual Display of Quantitative Information.* Cheshire, Conn: Graphics Press.

Wickham, H. (2007). Reshaping data with the reshape package. *Journal of Statistical Software* 21: 1–20. http://www.jstatsoft.org/v21/i12/.

Wickham, H. (2016). *ggplot2: Elegant Graphics for Data Analysis.* New York: Springer.

7

Working with Categorical Data

Categorical data or variables can only take on a finite number of distinct values. As defined in Chapter 1, we distinguish between three kinds of categorical variables: binomial, nominal, and ordinal. The definitions vary slightly between textbooks, and we will stick to the nomenclature used in Table 1.1.

Categorical data are common, just think of surveys ('strongly agree', 'agree', 'disagree', 'strongly disagree'), or observational studies in the natural sciences ('alive', 'dead'), for example. Of course, we have seen categorical variables in our earlier analyses, for instance where we had a treatment such as 'fertiliser' and a 'control'. There however, the categorical variables always represented predictors, and never response variables. In this chapter, we look at categorical variables in a broader sense, where they can be predictor or response variables, or both. The way we summarise, graph, and analyse such datasets is different from those with continuous response variables. In the real world, we often find ourselves with a mix of categorical and continuous predictor and response variables.

We will first focus on summarising, tabling, and visualising datasets, and then move on to look at ways of statistically analysing categorical data, as well as constructing predictive models.

Let us start with an example. We will simulate this dataset, so that you can easily reproduce it on your own.

```
> set.seed(68176916) # use this command to reproduce the same dataset
> surv <-  data.frame(survived = rbinom(132, 1, .4),
+                     genotype = round(runif(132, 0.5, 3.5)),
+                     method = round(runif(132, .8, 3.5)))
```

Let us assume we are studying a flea pest control method and have collected survival data on insecticide trials using three different application methods. Further, we have three different genotypes of fleas that we want to study our treatment on. So our variables will be 'survived' (binomial), 'genotype', and 'method', the latter two are both categorical, or, more specifically, nominal, i.e. they have several possible values that cannot be ordered quantitatively (see Table 1.1).

7.1 Tabling and Visualising Categorical Data

In order to get an overview of the dataset we just created, we can use familiar tools like `head`, `tail`, `summary`, and `str`.

R-ticulate: A Beginner's Guide to Data Analysis for Natural Scientists, First Edition.
Martin Bader and Sebastian Leuzinger.

Companion website: www.wiley.com/go/Bader

It is always good to know of what class your objects are (use `class` to check, see Box 1.4). Often, you can make simple changes by using `as.numeric`, `as.character`, `as.integer`, or `as.data.frame`.

```
> str(surv)
'data.frame':     132 obs. of  3 variables:
 $ survived: int  1 0 0 0 0 0 1 1 1 0 ...
 $ genotype: num  2 3 3 3 3 2 2 1 1 2 ...
 $ method  : num  3 3 2 3 1 3 1 1 3 2 ...
```

We can see that both 'genotype' and 'method' are currently variables of class `numeric`, which is a consequence of the way we simulated them using `runif`. We know that they are integers, so we may want to change this quickly. Also, 'survived' is an integer but should rather be of class `factor` (see Box 1.4).

```
> surv$genotype <- as.integer(surv$genotype)
> surv$method <- as.integer(surv$method)
> surv$survived <- as.factor(surv$survived)
```

In a dataset with continuous variables, we would want to see means, variances, and perhaps correlation metrics to quickly get an impression of how the variables are distributed, and how they stand in relation with each other. These metrics however make little sense for categorical datasets. A good first step here is a table of frequencies. For example, we can ask 'How many fleas survived given the three different application methods?'

```
> table(surv[, c('survived', 'method')])
        method
survived  1  2  3
       0 11 20 50
       1 18 20 13
```

It can be derived easily from this table that application method 1 works less well than method 2, which in turn works less well than method 3, where 50 out of 63 fleas do not survive. We can do the same thing using the variable 'genotype'. It is difficult to see both the effect of 'method' and 'genotype' on survival at the same time, as for this, our table would have to become three-dimensional. A proper analysis of the interaction between the two predictor variables and its effect on survival requires the use of a generalised linear model (see Chapter 12). However, there is a tool in R to visualise the relationship between more than two categorical variables, using the function `mosaicplot`.

We first plot two variables ('survived' and 'method') to confirm the insight gained from the table (Figure 7.1a).

```
> mosaicplot(survived ~ method, data = surv, main = '(a)  only method')
```

Indeed, method 3 seems much more effective than methods 1 and 2, as the area of the combination 0 (did not survive) and method 3 is much larger than 1 (did survive) and method 3. This is not the case for the other two methods (1 and 2). In Figure 7.1b, we can see the same plot, but the rectangles are divided up according to genotype.

```
> mosaicplot(survived ~ method + genotype, data = surv, main = '(b)
  method and genotype')
```

The interpretation of these plots becomes more complex the more variables you add. Figure 7.1b shows that for all three methods, the flea treatment works best on genotype 3, as those boxes are systematically larger than the ones for genotypes 1 and 2 on the left-hand side (where survival is 0). The surviving fleas seem to be predominantly from genotype 1, but for method 2, this shifts to genotype 2, hinting at a possible interaction between method and genotype.

The mosaic plot is clearly an exploratory plot rather than a plot used to present data. It gives us a feel for how categorical data correlate and interact.

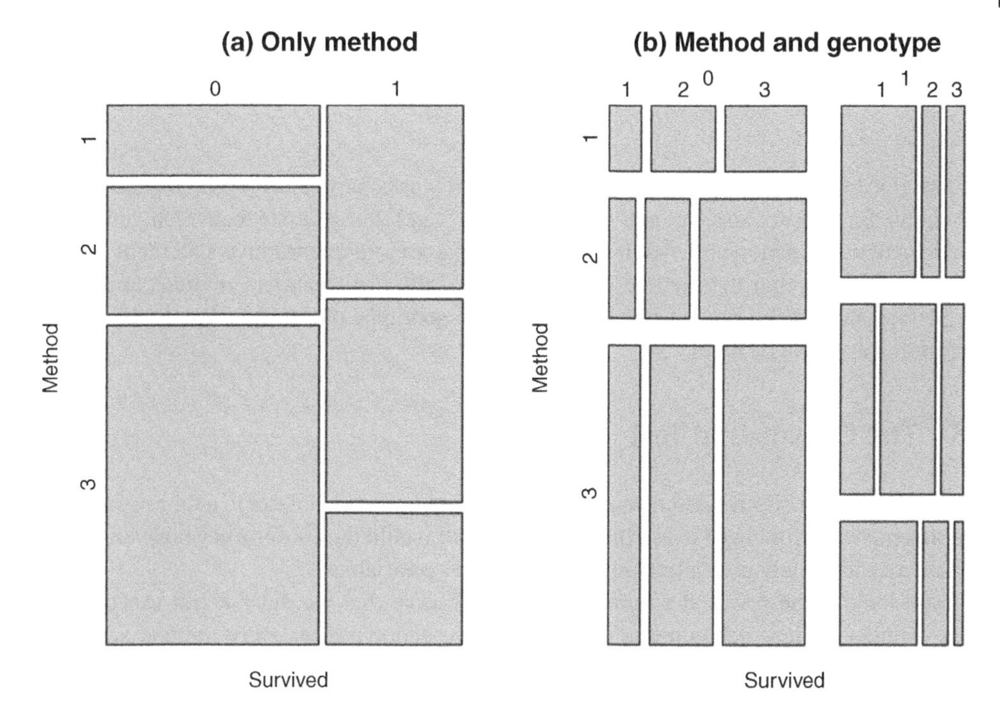

Figure 7.1 The mosaic plot (`mosaicplot`) provides a way of visualising categorical data. The relationship between two (a) or more (b) variables can be shown. In (b), the numbers 1, 2, and 3 across the top refer to the three genotypes.

7.2 Contingency Tables

The term 'contingency table' was coined by Karl Pearson and refers to the contingency (or dependency) of one variable on another

The table we obtained earlier by applying the `table` function can be expanded to what we call a *contingency table*. This historic term comes from the fact that the two categorical variables, if not independent, are *contingent* on one another. What is added are the *marginal sums* or *totals* as well as the *grand total*, the total number of entries in the dataset, here 132. Theoretically, a contingency table can include more than two categorical variables, but again, visualisation becomes difficult in three or more-dimensional space, so we will only look at two variables at a time.

```
> method <- table(surv[, c('survived', 'method')], dnn = c('survived',
   'method')) # dnn adds the variable names
> method <- addmargins(method) # addmargins adds the row and column totals
> method
        method
survived   1   2   3 Sum
     0    11  20  50  81
     1    18  20  13  51
     Sum  29  40  63 132
```

The contingency table tells us how many fleas died overall, and how many survived, but also how many times we used methods 1, 2, and 3 overall. We can produce the same table using the variables 'survival' and 'genotype':

```
> genotype <- table(surv[, c('survived', 'genotype')], dnn = c('survived',
   'genotype'))
> genotype <- addmargins(genotype)
> genotype
```

```
        genotype
survived   1   2   3 Sum
     0    14  27  40  81
     1    27  16   8  51
     Sum  41  43  48 132
```

From this table, we can see that the flea treatment works particularly well on genotype 3 (40 counts for 'did not survive' and only 8 for 'survived'), but is less effective on genotype 2 and particularly genotype 1. We now want to make a more quantitative statement about whether we think that the method of application of the flea treatment matters, and how the performance of the flea treatment relates to the genotype of the fleas. We do this next, performing a χ^2-test by hand.

7.3 The Chi-squared Test

To move from a qualitative description of what is going on in our dataset to a more quantitative statement, we first need to ask the question 'What would the contingency table look like if there was absolutely no relationship between the two variables?'

If you look at the row and column totals, we observe that we did not test exactly the same number of fleas belonging to any of the three genotypes for every method applied. Imagine for a moment that this was the case, so that we would have exactly 44 fleas for each genotype. Also imagine that out of the 132 fleas in total, exactly half died, half survived. In such a simple case (all row totals 44, both column totals 66), it would be easy to see that, if there was absolutely no dependency between the two variables, each of the six cases of the table would have to be 22 ($6 \times 22 = 132$). One way of calculating these expected counts is to multiply the row totals by the column totals and divide by the grand total ($44 \times 66/132 = 22$ for all six cases). Now that our column and row totals are in fact a little different, we can apply the same method of multiplying the row totals with the column totals, divided by the grand total. If we do this for all six cases, we obtain our *expected* or *theoretical* counts. In other words, the counts we would expect under our null hypothesis that *there is no dependency* between the two variables.

To calculate the expected or modelled counts for a contingency table, multiply each row total by the corresponding column total, then divide by the grand total

Once we have calculated our modelled counts, we can then look for a metric (a test statistic, see Chapters 3 and 5) that characterises the distribution of these counts, and ask: 'What would that test statistic look like if our counts were completely random, i.e. if there was no dependency between the two variables?'. Let us do this exercise by hand for the variables 'survived' and 'genotype'. We label the six cases according to 'survival' (0 or 1) and 'genotype' (1, 2, or 3) and add the suffix '_mod' to signify that these are our *modelled* (often also referred to as 'expected' or 'theoretical') counts if there were no dependencies.

The null hypothesis for a χ^2-test is: 'there is no dependency between the categorical variables (i.e. they are independent). The alternative hypothesis is thus: 'the two variables are related or dependent'

```
> sur0gen1_mod <- 41*81/132
> sur0gen2_mod <- 43*81/132
> sur0gen3_mod <- 48*81/132
> sur1gen1_mod <- 41*51/132
> sur1gen2_mod <- 43*51/132
> sur1gen3_mod <- 48*51/132
```

A way to capture the characteristic of a contingency table in a single number is to compare the modelled with the actual counts in each case. Pearson (1900) came up with the Chi-squared statistic, which does exactly that:

$$\chi^2 = \sum \frac{(x_{ik} - m_{ik})^2}{m_{ik}}$$

where the subscript i is the number of rows, k is the number of columns of the contingency table (excluding the row and column totals of course), x stands for the observed value, and m for the modelled value of the same case. If we calculate these six values and sum them up as per formula, we obtain:

```
> case1 <- (14 - sur0gen1_exp)^2/sur0gen1_mod
> case2 <- (27 - sur0gen2_exp)^2/sur0gen2_mod
> case3 <- (40 - sur0gen3_exp)^2/sur0gen3_mod
> case4 <- (27 - sur1gen1_exp)^2/sur1gen1_mod
> case5 <- (16 - sur1gen2_exp)^2/sur1gen2_mod
> case6 <- (8 - sur1gen3_exp)^2/sur1gen3_mod
> sum(case1, case2, case3, case4, case5, case6)
[1] 22.61941
```

So our test statistic is 22.62. The next step is, similar to what we have seen in Chapter 5 on the t-test, to compare this value to test statistics that would arise if we did the same calculation on random data. In other words, we ask 'How are such test statistics distributed if the two variables were independent?'. It turns out that they follow a so-called Chi-squared (χ^2) distribution with 2 degrees of freedom. Why two degrees of freedom? Imagine a blank 2×3 contingency table – how many counts can you choose freely once you have calculated the row and column totals? Try it out! It turns out exactly 2, after that, you are locked in to match the row and column totals.

A contingency table with i rows and k columns has $(i-1)(k-1)$ degrees of freedom, e.g. a 2×3 contingency table has $1 \times 2 = 2$ degrees of freedom

So we need to compare our χ^2-statistic of 22.62 to a χ^2-distribution with 2 degrees of freedom and ask 'Is a test statistic this extreme (or more extreme) common? If the answer to this question is 'yes', then we do not have sufficient evidence to reject the null hypothesis. If the statistic is sufficiently rare however, we reject our null hypothesis and turn to the alternative hypothesis. Let us test this by using the function pchisq, the analogous function to pnorm used in Chapter 3. Those functions calculate the probabilities associated with values of variables that follow a certain distribution (here the χ^2-distribution).

```
> pchisq(22.62, df = 2, lower.tail = F)
[1] 1.224981e-05
```

This tells us that to hit a χ^2-statistic of 22.62 or higher *by chance* comes with a very low probability of around 0.00001, which leads us to reject the null hypothesis (assuming a significance level of $\alpha = 0.05$). We can also visualise this by plotting our critical threshold of 22.62 onto a histogram of random numbers following a χ^2-distribution. This confirms that it would be extremely rare to observe such an extreme (or more extreme) value by chance (Figure 7.2).

Of course, R has a function to conduct the χ^2-test that we have just carried out by hand, called chisq.test. Its application is simple, and we can immediately see that the same critical value of 22.62 is shown, and the P-value corresponds to our probability calculation using pchisq.

```
> chisq.test(surv$survived, surv$genotype)

        Pearson's Chi-squared test

data:  surv$survived and surv$genotype
X-squared = 22.619, df = 2, p-value = 1.225e-05

> ## chisq.test(table(surv[, c('survived', 'genotype')])) yields the same
```

You can run the same test using the other variable, 'method', and you should be able to confirm that there is a significant dependency between 'survived' and 'method' as well.

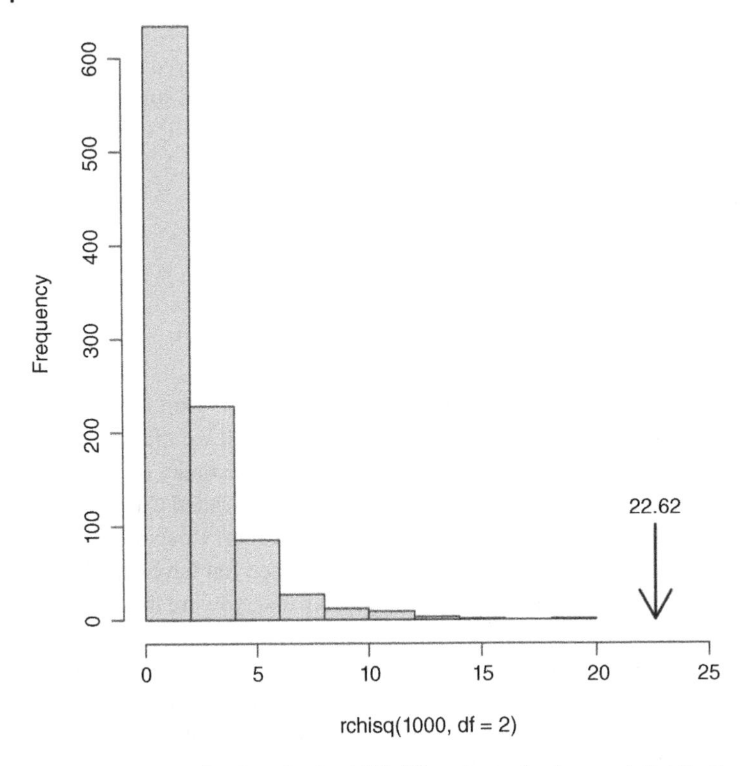

Figure 7.2 The χ^2-value obtained (22.62) against a background distribution of 1000 random numbers that follow a χ^2-distribution with 2 degrees of freedom. It can be seen that 22.62 or higher would be a rare draw from this pool of random numbers given this particular distribution, which lends us confidence to think that the effectiveness of our flea treatment method depends on the genotype of the fleas.

The χ^2-tests, together with the mosaic plots give us a good impression of how the effectiveness of our flea treatment (or their survival) depends on the application method and the genotype of the animals. What we are not able to do so far, is to predict if a certain animal will likely succumb to the treatment or not, given that it is of a certain genotype, and given a certain application method. For this, we need a predictive model, which we will introduce in the next section.

7.4 Decision Trees

Decision trees are intuitive, because we subconsciously use them all the time in our everyday decision-making. For example, we could weigh off the value of a trip to the countryside by first evaluating the weather (good or bad?) if good, we could then ask if there is a lot of traffic and make a final decision based on that (e.g. go or not even though traffic is heavy). Conversely, if the weather is bad, we could decide to go only if traffic is light. Decision trees are hugely important in machine learning, where complex decisions need to be made by automated algorithms. They work on datasets with both continuous and categorical variables. Fundamentally, they identify the optimal splitting points in a recursive manner, eventually categorising every observation of a dataset, mostly in respect to one response variable (e.g. the binomial decision 'go' or 'not go' in the simple example above). Large decision trees with many nodes are prone to overfitting, namely if too few observations end up in a given branch of the tree. Following Occam's razor (Chapter 1), we aim to find the simplest tree that explains the most, avoiding overfitting.

Decision trees are mostly used for predictive purposes rather than to test for statistical significance

Let us get back to our dataset from the previous section. To make things a little more interesting, we add a continuous variable. Imagine you had also measured the size of the tested fleas (in some arbitrary units).

```
> set.seed(31129) # so you can reproduce the dataset
> surv$size <- round(rnorm(132, mean = 20, sd = 3), digits = 1)
> head(surv, n = 3)
  survived genotype method size
1        1        2      3 15.5
2        0        3      3 21.3
3        0        3      2 22.5
```

We can now ask 'given the genotype, method used, and size of a flea, will it likely die or survive the treatment?' First, because we have added a continuous variable (size), we can have a quick look at a box plot to get an impression of whether bigger animals are more or less likely to respond to the treatment (Figure 7.3).

```
> boxplot(size ~ survived, data = surv)
```

If your dataset is very large, you may consider using a larger fraction to train the model, as your test dataset is still sufficiently large

Indeed it looks like size may have an influence on survival, so it could be useful to include this variable in our decision tree. In order to initiate the decision tree using `rpart` (Therneau and Atkinson 2023), we divide our dataset into two, one is our 'training' dataset with which we train the model, and one is our 'test' dataset, with which we test the performance of the model. It is important to keep the two separate, as we want an independent test of our predictive model. A common split into the two sets (training/test) is 70/30 or 80/20. It is important to select the data randomly, we use the function `sample` for this.

```
> set.seed(1) # setting a seed so you can reproduce this
> split <- sample(1:nrow(surv), 0.8*nrow(surv)) # 80/20 split
> trainSplit <- surv[split, ] # the training dataset
> testSplit <- surv[-split,] # the test dataset
```

We first run the model in its default mode:

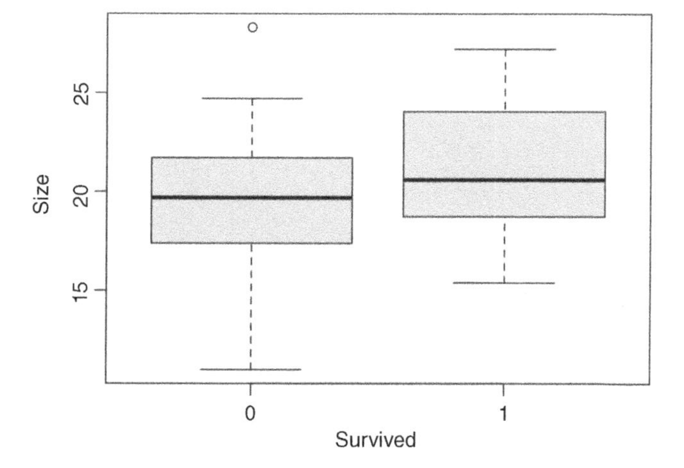

Figure 7.3 Survival of the fleas according to their size.

```
> library(rpart)
> m1 <- rpart(survived ~., data = trainSplit) # default rpart model
```

The tilde and the dot are to be read as 'predict survival using all available predictors'. Of course, we are using the training dataset for this. While you could look at the model output by typing 'm1' into the console, it is more informative to directly look at the plot of the decision tree (Figure 7.4).

```
> rpart.plot(m1)
```

There are a number of arguments you can play with, both for `rpart`, and for `rpart.plot`. The argument 'method' in `rpart` can be set if your response variable is categorical, continuous, or follows a Poisson or other distribution (time until death). The function is pretty smart and will usually make the right guess, but it might be more cautious to select a method up front, in our case: `rpart(survived ~., method = 'class', data = trainSplit)`. The other important argument is 'control', which we will look at later. The plotting function `rpart.plot` leaves quite a few options to display different metrics inside the node boxes, and a wealth of other graphical tweaks. For example, you can get rid of the fitted class (the top number in the boxes, see Figure 7.4) by choosing `rpart.plot(m1, extra = 107)`. The 'extra' argument is explained comprehensively in the help function (`?rpart.plot`).

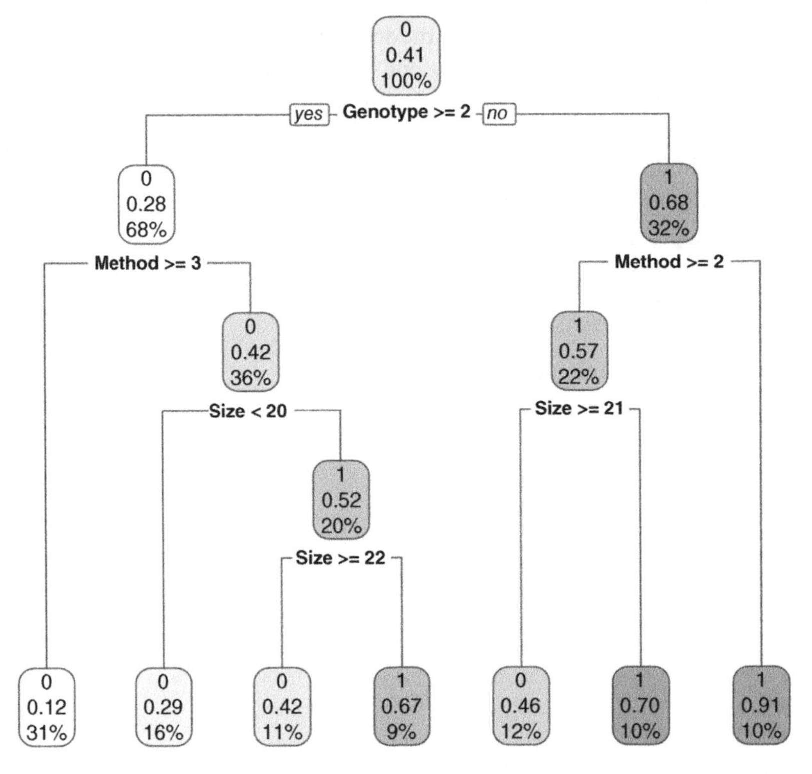

Figure 7.4 The result of a default plot produced by `rpart.plot`. The top numbers show if the entire cohort at that node will on average survive (1) or die (0). The numbers above the percentages are the probabilities of survival for the fleas at any given node (0.4 overall). The percentages indicate the proportion of individuals that fall into a node (100% in the first instance). The shading is proportional to survival (darker = high, brighter = low). Every node has a decision criterion (go left for 'yes', right for 'no'). Use the argument 'extra' to change the information displayed in the nodes.

Next, we of course want to see how well our model performs if we feed it the test dataset. Remember that the model has not seen these data, so this is a completely independent test. We can use the generic function `predict` to do this.

```
> predictTest <- predict(m1, testSplit, type = "class")
> predictTest
  4   5  11  16  27  36  46  47  49  52  56  57  58  60  61  62  67  77
  0   0   0   1   0   0   0   1   0   0   0   0   0   0   1   0   0   0
 80  83  88  94 104 111 118 121 127
  0   0   1   1   0   0   1   1   1
```

The accuracy of a decision tree is the correct classifications over the total classifications

We can see the prediction for every line of the test dataset. To easily see how many classifications the model got right (the accuracy of the model), we can use the function `table`, which summarises the matching zeros (flea died) and ones (flea survived).

```
> table(predictTest, testSplit$survived)

predictTest  0  1
          0 16  3
          1  3  5
```

So out of the nineteen fleas that did not survive, 16 were correctly predicted to die, but 3 were wrongly predicted to die. Conversely, 5 were correctly predicted to survive, but 3 were wrongly predicted to die (they survived). You could also say that the model did well in 78% of all cases, but failed 22% of the time, simply by dividing the correct classification (21) by all cases (27).

7.5 Optimising Decision Trees

There is a risk of under- or overfitting a decision tree. Underfitting means that we could have done better with more nodes, and overfitting means that the tree is too complex (too many nodes). To optimise a decision tree, we have a few options. If you look at the help file of `rpart.control`, you will see that we can play with quite a number of buttons and levers, the most important being the complexity parameter ('cp'), the minimum number of observations per node ('minsplit'), the minimum number of observations at the end of each branch, or the leaf ('minbucket'), and the number of cross validations ('xval'). Cross validation is an algorithm that iteratively uses only part of the dataset and validates with the remaining part. There are no standard or widely accepted protocols on how to arrive at the optimal tree, but it is worth learning some basic tweaks. To quickly evaluate what happens to the power of our model, we programme a very simple function (see Box 1.1) that returns the accuracy (correct classifications over total classifications), just like we have calculated it above.

```
## accuracy function
accuracy <- function(x, y) { # x and y are the modelled and true
    classifications
  round(100 * sum(diag(table(x, y)))/sum(table(x, y)))
}
```

We can quickly test our function and see if we can confirm the 78% accuracy of our default model (`m1`).

```
> accuracy(predictTest, testSplit$survived)
[1] 78
```

That worked. Let us look at the default parameter settings of `rpart.control` and start tweaking things from there. The help file of `rpart.control` will show you this:

```
rpart.control(minsplit = 20, minbucket = round(minsplit/3), cp = 0.01,
   maxcompete = 4, maxsurrogate = 5, usesurrogate = 2, xval = 10,
   surrogatestyle = 0, maxdepth = 30)
```

First, let us reduce the number of minimum observations per node to 1 (which means no restrictions on the minimum number) and call this model 2 (m2).

```
> m2 <- rpart(survived ~., data = trainSplit, control =
   rpart.control(minsplit = 1))
> predictTest <- predict(m2, testSplit, type="class")
> accuracy(predictTest, testSplit$survived)
[1] 63
> rpart.plot(m2) # figure not shown
```

The resulting tree is massively overfitted, but not only that, we also lost accuracy. So this was not a good idea. You will see that as you increase 'minsplit' slightly (to 5 for example), you will quickly reach our previous accuracy of 78%, but the tree is still overfitted. This is because the resulting tree (look at the figure obtained via `rpart.plot`) has many more nodes than our original tree, and we know that at equal explanatory power, 'simpler is better' (cf. Occam's razor, Chapter 1). In fact, you can increase 'minsplit' all the way to 38 and not lose accuracy (but nodes!). So for now, `m1` is winning over `m2`.

Next, we look at the complexity parameter, which by default is set to 0.01. The parameter indicates a minimum improvement (prediction power) that needs to be made at any given node for the node to be justified. Therefore, if you set it to a low value (0.001 would be considered very low), your model will show a lot of nodes, and if you set it high (0.1), your model will be very conservative and avoid nodes that do not explain much, leading to a small tree with only few nodes. For bigger datasets with more nodes, you can use `plotcp`, to help locate an optimal complexity parameter. In the case of our small dataset, we can just probe around the default value of 0.01 to see if this gives us better accuracy. You will note that when you reduce the complexity parameter, not much will happen, as 'minsplit' will prevent nodes from forming. However, if you increase 'cp' to say 0.02 (leaving 'minsplit' at the default value of 20), you will observe that one node will be eliminated, and the classification improves slightly to an accuracy of 81%:

Decision trees can be used as an alternative to linear regression

```
> m3 <- rpart(survived ~., data = trainSplit,
+   control = rpart.control(minsplit = 20, cp = .02))
> predictTest <- predict(m3, testSplit, type = "class")
> accuracy(predictTest, testSplit$survived)
[1] 81
> rpart.plot(m3)
```

so that overall, `m3` will be the preferred model. It is important to understand that due to the small number of observations in this dataset (particularly in the test dataset), things behave rather stochastically. The cross validations (using random training and test datasets during the model fit) for example are different every time you fit the model, making it worthwhile to play a little bit with some of the `rpart.control` parameters in order to see how stable the model fit is. Another approach is to deliberately overfit the model at the start, and then 'prune' the tree using `prune`. Finally, remember that decision trees can also be used to predict continuous variables. In our case, for example, you could use `rpart` to predict flea size, given the method of application, their survival, and their genotype. As such, decision trees are powerful non-parametric alternatives to linear models that are used to predict a given response variable.

References

Pearson, K. (1900). On the criterion that a given system of deviations from the probable in the case of a correlated system of variables is such that it can be reasonably supposed to have arisen from random sampling. *Philosophical Magazine Series 5* 50 (302): 157–175. https://doi.org/10.1080/14786440009463897.

Therneau, T. and Atkinson, B. (2023). rpart: recursive partitioning and regression trees. R package version 4.1.21. https://CRAN.R-project.org/package=rpart.

8

Working with Continuous Data

In Chapter 1 we have seen different types of variables and learnt how to confidently distinguish between them. Continuous variables, or, more broadly, continuous data contain numeric values that can take on any real number. Examples were given in Chapter 1, but to quickly remind ourselves – 'body height' for instance is a continuous variable, as it can take on any number with infinite decimals (within reasonable boundaries of course). Once we know that we are dealing with continuous variables, we need to know (i) if they will act as predictor or response variables in our dataset (Chapter 1), and (ii) what distribution they resemble (Chapter 3). This will inform our subsequent analysis. Box 12.2 can be used to help determine what type of analysis will give us the answers we are looking for.

In this chapter, we will cover some basic concepts that relate to continuous data and then have a detailed look at correlation analysis. Just like in the other chapters, we generally assume that our data follow a normal distribution (at least approximately). Because the real world shows us that this is often not the case, we have three options: (i) transform the data hoping they will then meet the assumption, (ii) use a so-called 'non-parametric', rank-based method, or (iii) turn to entirely different, more modern methods that can account for non-normality. Option (iii) has become by far the most common, as it leaves us with maximum flexibility. Nevertheless, we will briefly treat the most common data transformations here, as well as point you to the classic non-parametric methods, as we have done in Chapter 5.

8.1 Covariance

Building on the concept of the *variance* (Chapter 2), we can define a *covariance* when looking at two continuous variables concurrently. The question we are asking is 'do the variables vary in a similar way?' We extend the definition of the variance

$$\frac{\sum (x_i - \bar{x})^2}{n - 1}$$

and replace $(x_i - \bar{x})^2$ by $(x_i - \bar{x})\,(y_i - \bar{y})$ to take into account a second variable y. The denominator remains, reflecting the number of values in x or y (which must be the same), minus 1. See Section 2.6 for a brief explanation why we divide by $n - 1$ rather than by n when dealing

R-ticulate: A Beginner's Guide to Data Analysis for Natural Scientists, First Edition.
Martin Bader and Sebastian Leuzinger.

Companion website: www.wiley.com/go/Bader

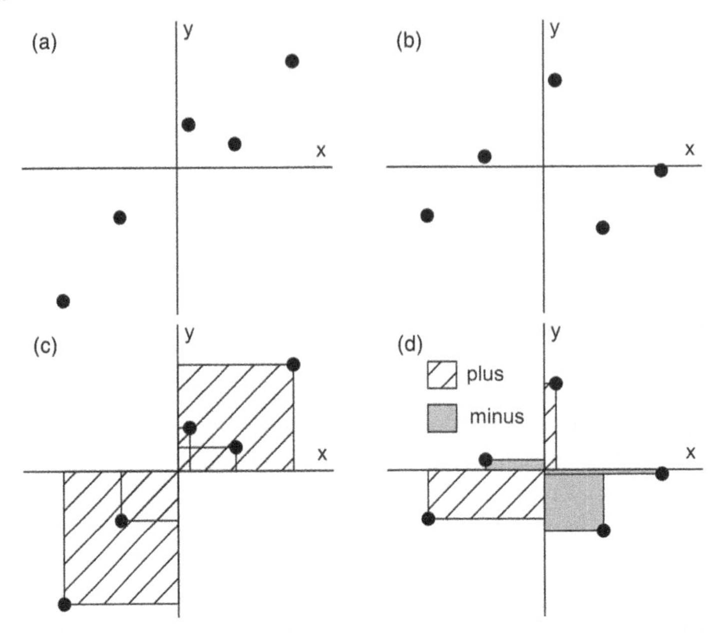

Figure 8.1 Visualising the concept of covariance. In (a), x and y appear to be correlated, while in (b), they are not. When calculating the covariance, the products of the values of x and y are positive (hatched rectangles) if x and y are positive or negative at the same time (c). In (d), that is not the case – some products are positive, and some are negative (grey rectangles), leading to a sum of products that is relatively close to zero indicating that x and y are not correlated.

with a sample. We therefore define *covariance* as follows:

$$\text{COV}(x, y) = \frac{\sum (x_i - \bar{x})(y_i - \bar{y})}{n - 1}$$

So what exactly does covariance tell us? In brief, it becomes large when the two variables 'co-vary' in a similar way, i.e. if the values of x become positive or negative at the same time as the values of y become positive or negative. This is shown in Figure 8.1. Because we are adding the products of corresponding x and y values (see the above formula), the sum, and therefore the covariance, becomes large (in absolute terms) when the pairs vary in a similar fashion (as the product of two negative numbers is positive, see Figure 8.1a,c). If the corresponding x and y values vary randomly, the products of x and y are sometimes positive, and sometimes negative, leading to a sum of products that is close to zero, meaning that x and y are little or not correlated (Figure 8.1b,d). Note that in a negative correlation, the sum of products will be negative, but large in absolute terms (i.e. you would see a lot of grey rectangles in Figure 8.1).

8.2 Correlation Coefficient

So if the covariance is large (in absolute terms – it could also be negative), this tells us that the two variables involved are correlated, if it is close to zero, they are not. But how close to zero means no correlation? And what is large? To answer these questions, we introduce the 'correlation coefficient', which standardises the covariance, such that we end up with a convenient metric that is comparable between datasets: from −1 (a perfect negative correlation) to zero (no correlation whatsoever), and +1 (a perfect positive correlation). We will not delve deeper into the mathematics here, but simply acknowledge that we obtain such a standardised metric by dividing the covariance by the product of the standard deviations of x and y:

Covariance can take on values from $-\infty$ to $+\infty$, while the correlation coefficient ranges from −1 to +1.

$$r = \frac{\text{COV}(x,y)}{s_x\, s_y} = \frac{\sum (x_i - \bar{x})(y_i - \bar{y})}{\sqrt{\sum (x_i - \bar{x})^2 \sum (y_i - \bar{y})^2}}$$

This formula represents the 'Pearson' correlation coefficient. It is the most common way of characterising how two approximately normally distributed variables are related. Learning to interpret correlation coefficients is an important skill, and expressing them as 'percentage of shared variance' will help. For instance, a correlation coefficient of 0.2 (or −0.2) between x and y will read: '20% of the variance in x is shared with y', and a correlation coefficient of +/−0.95 '95% of the variance in x is shared with y'. But what value of r testifies a weak, medium, or strong correlation? While sometimes correlations >0.5 are considered 'strong', the interpretation depends on the subject area and the specifics of each case. In chemistry, 20% 'shared variance' (i.e. a correlation coefficient of 0.2) might be considered weak, but it may be thought of as substantial in a subject like psychology or ecology, where generally we deal with low signal-to-noise ratios. Figure 8.2 illustrates how correlation strength and correlation coefficient are related.

To compute a correlation coefficient in R, we use the function `cor`:

```
> cor(1:5, 1:5) # a perfect positive correlation (r = 1)
[1] 1
> cor(1:5, c(2, 1, 4, 7, 7)) # a tight positive correlation
   (r = 0.91)
[1] 0.9116846
```

It is again important to note that the validity of these correlation coefficients relies on a critical assumption that we have encountered earlier: we require that our variables in question (at least approximately) follow a normal distribution. How to test for normality using both visual and numerical methods has been discussed earlier (see Section 5.2.1.1). So what if the assumption of normality is violated, i.e. we are facing a set of non-normal variables? One option is to transform the original variables, we will treat this in Section 8.3. Another option is to use a non-parametric, or 'parameter-free' method to estimate the degree to which two variables are correlated. Two metrics exist that work on ranks instead of the absolute values of our two variables in question: Spearman's and Kendall's rank correlation coefficients. Both

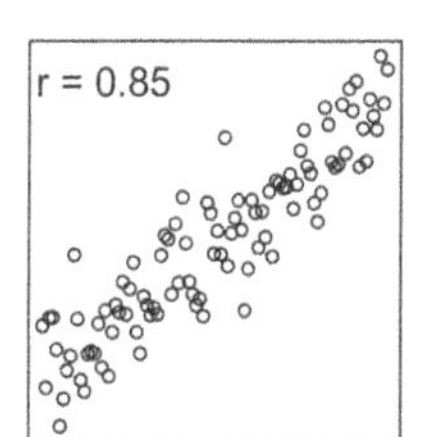

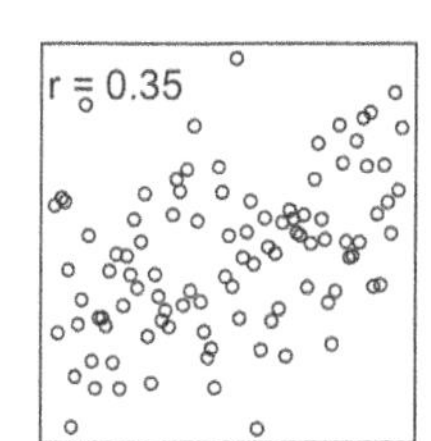

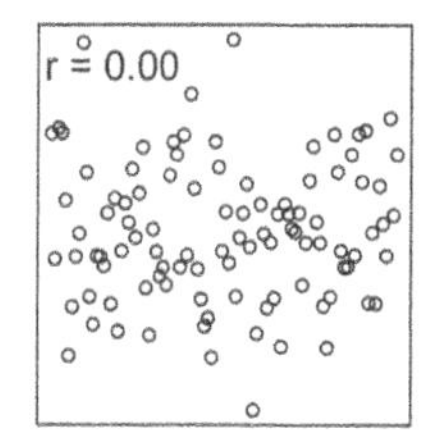

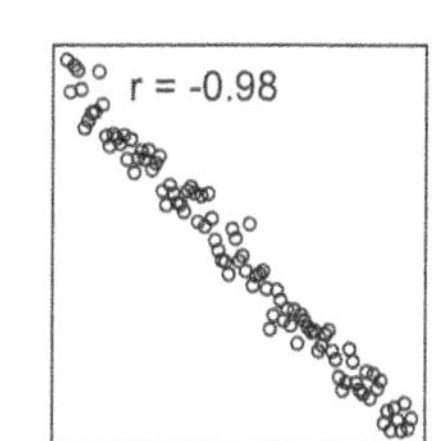

Figure 8.2 A visualisation of some typical correlation coefficients. It is important to develop a sense for what a given correlation coefficient approximately looks like when the data are plotted.

Non-parametric tests are tests that do not rely on the data to follow certain distributions, which again rely on parameters.

methods are very closely related, and we will focus on Kendall's rank correlation coefficient τ ('tau') here, which is often preferred as the slightly more robust method (Kendall 1938). The coefficient τ quantifies the similarity of the rankings of the paired observations. We will not look deeper into the mechanics of how τ is calculated, for now it is important to know how to specify such a non-parametric correlation coefficient in *R*. We can use the same variables as above, in fact, it would have been reasonable to use the non-parametric alternative in the first instance for the correlation coefficient, simply because verifying the assumption of normality is nearly impossible with such small datasets:

```
> cor(c(1:5), c(1:5), method = 'kendall') # a perfect posi-
tive correlation [1] 1
> cor(c(1:5), c(2, 1, 4, 7, 7), method = 'kendall') # r is now
   much lower!
[1] 0.7378648
```

As expected, the first pair of perfectly correlated variables still receives a correlation coefficient of 1. The second pair, which resulted in an r of 0.91 using the standard method, now shows a considerably lower correlation coefficient τ of 0.74. In the section on correlation tests, we will show that if we fail to test for normality, we risk inflating our type I and type II errors. This means that not only will we get different correlation coefficients if we ignore the non-normality of our data, but also, and worse, we can no longer trust our *P*-values (see Chapter 3).

8.3 Transformations

Transformations used to be important before more sophisticated methods that are able to relax the assumption of normality were developed (such as the generalised linear model, commonly known as GLM). While a lot less common now, data transformations are still useful for example to better visualise data. Performing and understanding the concept of transforming (and back-transforming) data is also an essential mathematical tool that you need to navigate the jungle of statistical methods. We will first show some very common transformations, but then quickly turn to the extremely powerful `boxcox` function in R, which is a bit like the Swiss army knife of transformations.

The square root and the log transformations both compress large values, the latter does so a little more aggressively. Classic applications for these transformations are variables like income or house prices, where a few extreme values distort the distribution. Figure 8.3 shows what transforming a left-skewed variable first with the square-root, and then with a log transformation can look like. The figure shows histograms and quantile-quantile-plots (Q-Q plots) of the untransformed and transformed variable. Note that we are using *ggplot2* here. You have already seen how to draw histograms and Q-Q plots using base *R*, here you can see the power of using *ggplot2*, for example by adding a density curve to the histogram with ease.

```
> x <- rchisq(200, df = 6) # create some non-normal data
> d1 <- data.frame(x = x, x1 = sqrt(x), x2 = log(x))
> ## Histogram with density plot
> p1 <- ggplot(d1, aes(x = x)) + geom_histogram
   (aes(y = after_stat(density)))
```

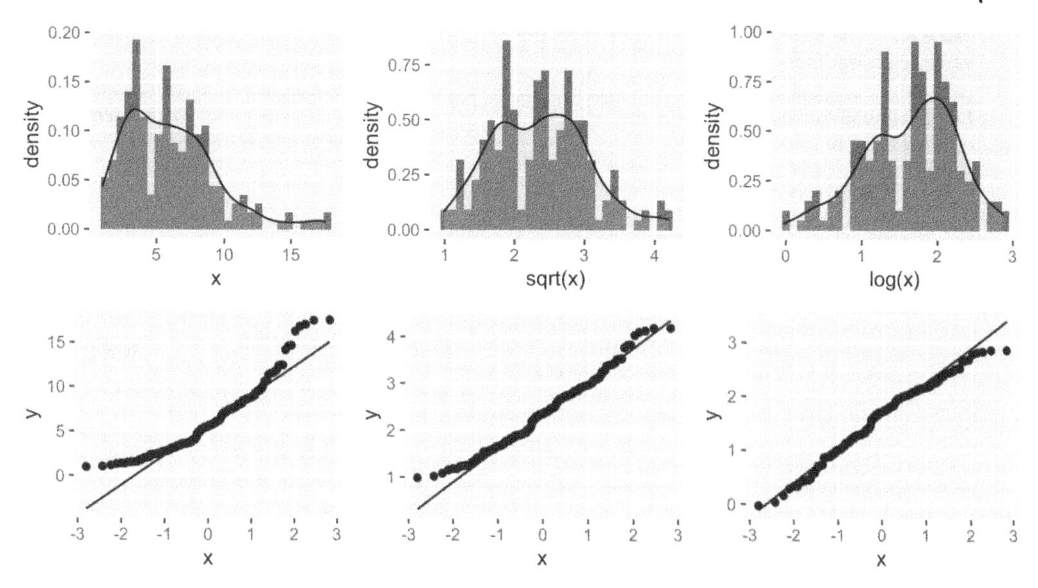

Figure 8.3 The square root and log transformations. Raw (untransformed) data on the left, square root transformation in the middle, and log transformed data shown on the right-hand side. Top row shows histograms including a density curve, and bottom plots show qq-norm plots, i.e. the quantiles of the sample distributions plotted against the quantiles of the standard normal distribution. The closer the data points to the line, the more the sample distribution resembles a normal distribution.

```
+ geom_density(alpha = .2, fill = "grey")
> p2 <- ggplot(d1, aes(x = sqrt(x))) + geom_histogram
  (aes(y = after_stat(density))) +
+ geom_density(alpha = .2, fill = "grey")
> p3 <- ggplot(d1, aes(x = log(x))) + geom_histogram
  (aes(y = after_stat(density))) +
+ geom_density(alpha = .2, fill = "grey")
> ## Q-Q plots:
> p4 <- ggplot(d1, aes(sample = x)) + stat_qq() + stat_qq_line()
> p5 <- ggplot(d1, aes(sample = sqrt(x))) + stat_qq() +
  stat_qq_line()
> p6 <- ggplot(d1, aes(sample = log(x))) + stat_qq() +
  stat_qq_line()
> grid.arrange(p1, p2, p3, p4, p5, p6, nrow = 2) # arrange in a
  2x3 grid
```

The `boxcox` function lets you quickly find the most effective transformation to render your data more normal

It appears that the square root and log transformations succeed in rendering the data more normal, note for example how the left-hand tail of the Q-Q plots becomes more aligned going from left to right, this is less apparent on the right-hand end of the distribution. So which transformation out of the two should be preferred in this case if the goal is for the data to become more normal? Judging from the Q-Q plots, this does not become entirely clear. A really helpful tool to determine which transformation is best suited to get closer to normality is the 'Box-Cox' transformation. In R, we use the `boxcox` function, which we find inside the built-in package MASS (Venables and Ripley 2002). Table 8.1 shows us the mathematics behind this – put simply, we look for the value of λ that maximises the expression

Table 8.1 The Box-Cox transformation finds the optimal λ value that should be used to make a variable appear more normal.

Log-likelihood function for finding the optimal value of λ	λ	**Transformation**
$\frac{x^\lambda - 1}{\lambda}$ if $\lambda \neq 0$ or	−2	$\frac{1}{x^2}$
$\log(x)$ if $\lambda = 0$	−1	$\frac{1}{x}$
	−0.5	$\frac{1}{\sqrt{x}}$
	0	$\log(x)$
	0.5	$\sqrt{x}$
	1	x
	2	x^2

on the left side of Table 8.1, which informs us about the most efficient transformation. For example, if the maximum is found at λ close to 0.5, we should use a square-root transformation. Let us see what the `boxcox` function suggests in our example from earlier. The function requires the output of a linear model to be supplied, so in the simplest case, we fit an `lm` of the data in question against their index. For more details on the use of `lm`, refer to Chapter 9.

```
> library(MASS) # load the necessary package
> m1 <- lm(x ~ 1)
> b1 <- boxcox(m1) # produces a figure
>
> ## If needed, the exact lambda can be extracted like this:
> lambda <- b1$x[b1$y == max(b1$y)]
> lambda
[1] 0.7070707
```

To determine the best transformation, you can either look at the standard plot that is shown when you call the `boxcox` function (not shown here, but try it out). Choose the λ that is closest to the maximum log-likelihood. In our case, this is likely 0.5. If unsure, you can compute the exact value of the optimal λ as shown in the code earlier. In fact, 0.7 is closest to 0.5, so a square-root transformation is most useful here (Table 8.1). You could of course transform your raw data with the optimal λ value (here 0.7) and the formula given in Table 8.1 instead of matching it to the closest conventional transformation. However, in most cases, using one of the classic transformations will be nearly as powerful and a bit easier to perform.

8.4 Plotting Correlations

For plotting correlations, you can follow the basic instructions in Section 6.2 on scatter plots. Here, we will add two additional tools that can be useful in plotting continuous variables. First, it is sometimes useful not to plot on the original scale(s), but on the transformed one(s). The reason for this is simply readability – a few large values in a dataset can require you to set large limits for the x- and y-axes, 'squashing' the majority of the data into a very

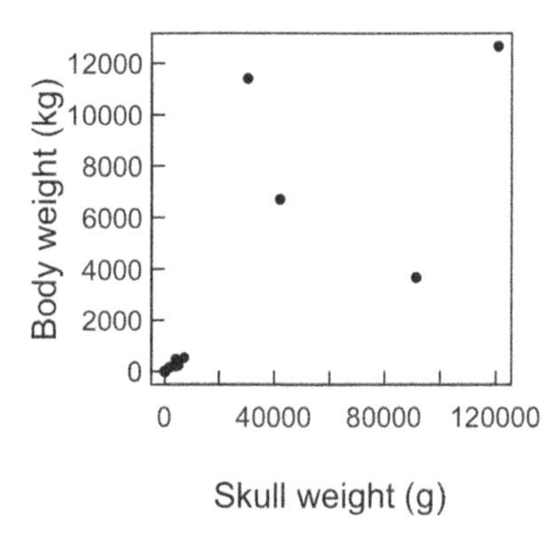

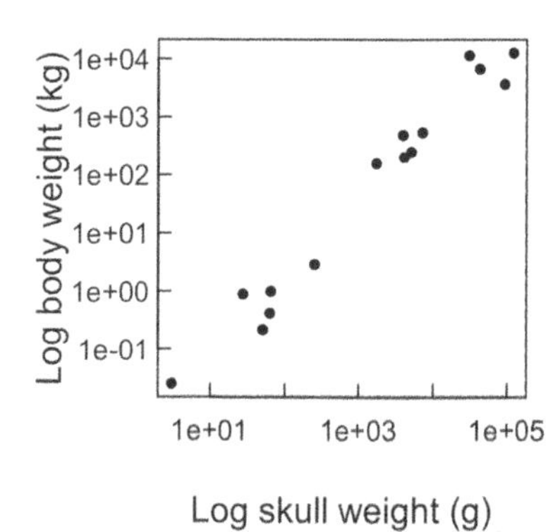

Figure 8.4 A scatter plot of skull weight vs. body weight of a number of animals. The log transformation of both axes makes the graph visually a lot more accessible. The transformation is also useful before applying a standard correlation test, in order to meet the assumption of normality (see Section 8.5).

small space, making the graph hard to read. We will illustrate this using a made up dataset on body weight vs. skull weight of various animals (Figure 8.4).

Another important tool, mostly used for data exploration, is the 'pairs' plot, in which all possible correlations within a dataset (mostly with continuous data) are shown simultaneously. While `pairs` works well for a quick exploratory plot (you can simply provide the entire dataset as an argument), we will show you a more elaborate version using the package *psych* (Revelle 2023). We will use the 'swiss' dataset, that is built into R:

```
> head(swiss, 3)
             Fertility  Agriculture  Examination  Education  Catholic  Infant.
                                                                       Mortality
Courtelary        80.2         17.0           15         12      9.96     22.2
Delemont          83.1         45.1            6          9     84.84     22.2
Franches-Mnt      92.5         39.7            5          5     93.40     20.2
```

The dataset contains six continuous variables that have been collected for 47 villages in Switzerland in the 19th century. Use `?swiss` to learn more about the dataset.

```
> library(psych) # you have to download and install the package first
> pairs.panels(swiss,
+              method = "pearson", # correlation method
+              hist.col = "grey", # colour of histograms
+              density = TRUE,  # show density plots
+              ellipses = F, scale = T, cex.cor = 2.5 # additional settings
+ )
```

Pairs plots are mainly used as an exploratory tool for datasets with multiple continuous variables

The resulting plot is rather powerful, showing all distributions of individual variables using histograms, density plots, pairwise correlations with spline fits, and all correlation coefficients printed proportional to their absolute magnitude (see Figure 8.5). This allows quick interpretations of how the variables are distributed and whether they are likely normal, which ones are correlated and to what degree, and whether the relationships are likely linear or not. If you would like to explore other variants of this type of plots, including some that can incorporate a grouping variable, see the `ggpairs` function in package *GGally*, an extension of the *ggplot2* universe (Schloerke et al. 2021).

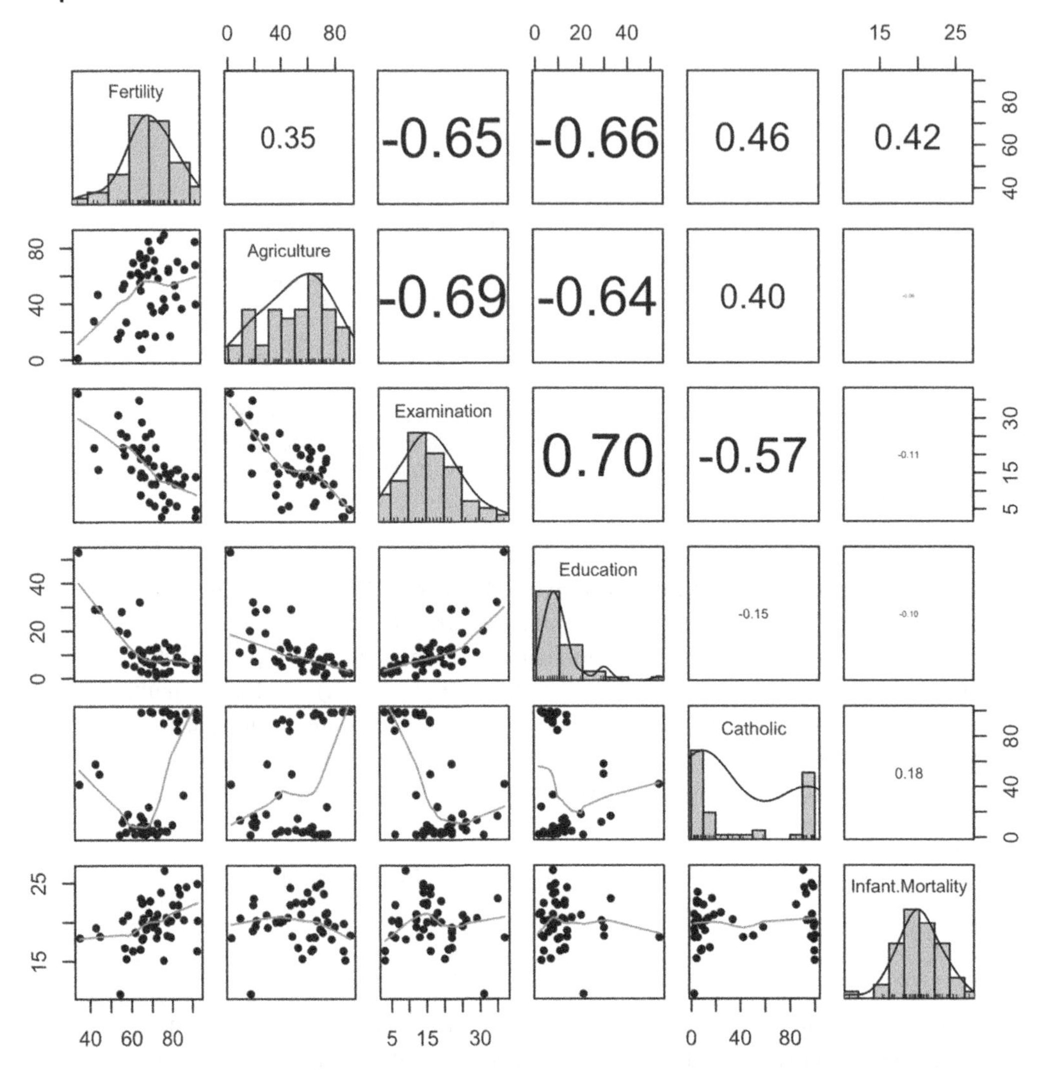

Figure 8.5 An embellished 'pairs' plot using the package *psych* and its function `pairs.panels`. All plots in the leftmost column feature the variable 'fertility' on the x-axis. In the second column, 'agriculture' is on the x-axis, and in the third column 'examination' occupies the x-axis, and so forth. The top right shows the correlation coefficients, with the font size proportional to the strength of the correlation. The diagonal shows histograms with density curves and 'carpets' (tiny black vertical lines) on the x-axes, indicating where the data sit. The bottom left half shows pairwise scatterplots, with a spline function (see Chapter 12) fitted to indicate the nature of the correlation.

8.5 Correlation Tests

Ultimately, we also want to test whether correlations are significant, using traditional frequentist arguments. The null hypothesis will be 'there is no significant correlation between the two variables'. The test statistic that will be used is the t-statistic, the same as is used for the t-test (Chapter 5). The test is easily conducted in R using `cor.test`. We could for example test for the (rather obvious) correlation between the variables 'agriculture' and 'examination' in the `swiss` dataset used earlier:

```
> cor.test(x = swiss$Agriculture, y = swiss$Examination)

        Pearson's product-moment correlation

data:  swiss$Agriculture and swiss$Examination
t = -6.3341, df = 45, p-value = 9.952e-08
alternative hypothesis: true correlation is not equal to 0
95 percent confidence interval:
 -0.8133545 -0.4974484
sample estimates:
       cor
-0.6865422
```

The output shows us which type of correlation was used (in this case the standard 'Pearson' correlation), what data were used, the *t*-value, the degrees of freedom (the sample size minus 2), the *P*-value, the alternative hypothesis, and the correlation coefficient including a confidence interval. Depending on whether we consider our two variables to follow a normal distribution, we can use the 'Kendall' or 'Spearman' (non-parametric) methods as discussed in Section 8.2. If in doubt, it is more cautious to use a distribution-free test that is based on ranks. Consider this example:

```
> cor.test(x = swiss$Catholic, y = swiss$Agriculture, method =
   'pearson')

        Pearson's product-moment correlation

data:  swiss$Catholic and swiss$Agriculture
t = 2.9372, df = 45, p-value = 0.005204
alternative hypothesis: true correlation is not equal to 0
95 percent confidence interval:
 0.1287588 0.6171749
sample estimates:
      cor
0.4010951

> cor.test(swiss$Catholic, swiss$Agriculture, method =
   'spearman')

        Spearman's rank correlation rho

data:  swiss$Catholic and swiss$Agriculture
S = 12303, p-value = 0.04907
alternative hypothesis: true rho is not equal to 0
sample estimates:
      rho
0.2886878
## there will be a warning message, which you can safely ignore
```

For sample sizes less than about 30, use non-parametric (rank-based) correlation tests

The correlation between the variables 'catholic' and 'agriculture' is clearly significant if we apply a standard correlation test, using the 'Pearson' method. If we use a rank-based method though (here the 'Spearman' rank correlation test is shown), the *P*-value is only just scraping the 5% significance threshold. Looking at Figure 8.5, we can ascertain that neither of the two variables look normal, so we should definitely trust the rank-based, non-parametric method more. You can easily see how we can walk into the trap of inflated type I but also type II errors if we fail to carefully assess the normality assumption. Also, interpreting a (significant) *P*-value should as always be carried out with caution: we know that the 5% threshold is arbitrary for example, and there is no 'law' or compelling reason why it should sit at exactly 5% (see Chapter 5).

In conclusion, correlation analysis should be viewed as a simple way of quickly detecting possible relationships between variables, and to get an overview over large datasets. The non-parametric alternatives give some flexibility for when variables are non-normal, and/or are not related in a linear fashion. Often however, the journey will not stop at a correlation analysis, but it will inform the next steps which could include a more complex regression model, or a multivariate analysis for example.

References

Kendall, M.G. (1938). A new measure of rank correlation. *Biometrika* 30: 81–93. https://doi.org/10.1093/biomet/30.1-2.81.

Revelle, W. (2023). Psych: procedures for psychological, psychometric, and personality research. Northwestern University, Evanston, Illinois. R package version 2.3.6.

Schloerke, B., Cook, D., Larmarange, J. et al. (2021). GGally: Extension to 'ggplot2'. R package version 2.1.2.

Venables, W.N. and Ripley, B.D. (2002). *Modern Applied Statistics with S*, 4th ed. New York: Springer. ISBN 0-387-95457-0.

9

Linear Regression

9.1 Basics and Simple Linear Regression

The idea behind simple linear regression is based on a systematic relationship between two continuous variables, where the value of one variable (the response) depends on the value of the other variable (the predictor or explanatory variable). This is the key difference to correlation analysis treated in Chapter 8, where the two continuous variables are interchangeable. Just like in correlation analysis, the relationship can be positive (the larger the predictor value, the larger the response value) or negative (the larger the predictor value, the lower the response value). The linear relationship is determined by two parameters:

- the intercept (β_0) – the value of the response variable (y) when the explanatory variable (x) is zero.
- the slope (β_1) – the rate of change in y associated with a one-unit change in x.

The term 'regression' was first used by a doctor who plotted body height of parents and their offspring. He noted that the metric 'regressed' to the centre (mean) – very tall fathers had shorter sons, very short fathers taller sons.

Together with the *error term* ε (also known as *residual error* or *random error*), representing real-world variation, the intercept and slope form the basic linear regression model:

$$y_i = \beta_0 + \beta_1 x_i + \varepsilon_i \quad \varepsilon \sim N(0, \sigma^2)$$

where the response variable is indicated by y, and x represents the explanatory variable. The subscript i denotes the running index and refers to an individual observation (that means i runs from the first to the last observation as in 1 … n). The $\varepsilon \sim N(0, \sigma^2)$ bit describes the assumption that the model errors (estimated by the residuals) follow a normal distribution with a mean of zero and variance σ^2. Figuratively speaking, this means that our observations scatter randomly around the straight regression line with a certain, relatively constant variance over the entire predictor range. Note that in this simple case, errors are only accounted for in the response space (on the y-axis), and we assume that the predictor values (on the x-axis) are measured without error.

Linear regression models rely on a technique called *Ordinary Least Squares*, which aims at finding intercept and slope values that minimise the sum of the squared differences (residuals) between the observed and predicted response values. This quantity is called the *residual sum of squares* (**RSS**). So, a residual (ϵ_i) is simply the difference between an observed response value (y_i) and the corresponding fitted value ($\hat{y}_i$) computed by the model (Figure 9.1):

$$\varepsilon_i = y_i - \hat{y}_i$$

$$\text{RSS} = \sum_{i=1}^{n} \varepsilon_i^2$$

R-ticulate: A Beginner's Guide to Data Analysis for Natural Scientists, First Edition.
Martin Bader and Sebastian Leuzinger.

Companion website: www.wiley.com/go/Bader

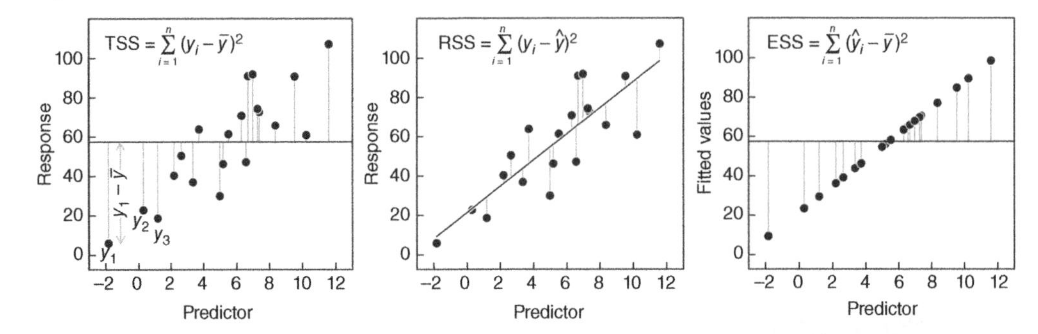

Figure 9.1 Total, residual, and explained sum of squares (***TSS***, ***RSS***, ***ESS***). An individual residual is represented by a grey vertical line. y_i = ith observation (of the response variable, $n = 20$), $\bar{y}$ = overall mean of the response variable (indicated by the horizontal line in the left and right panels), $\hat{y}_i$ = fitted value (predicted value) of the ith observation. You can see that TSS (left) is the sum of RSS and ESS (two graphs on the right).

The residuals are the differences between observed and predicted response values. They represent the unexplained variation that the model could not account for.

The residuals represent the proportion of the variation in the response variable that could not be explained by the model and are thus the best estimate of the error of the linear model.

The total variation seen in the data is estimated by the total sum of squares (**TSS**), which refers to the squared deviations of the observations from the overall mean (Figure 9.1).

The explained sum of squares (**ESS**) are simply the squared deviations of the fitted values (model predictions) from the overall mean, such that TSS = ESS + RSS (Figure 9.1).

We will use the dataset `leaf_respiration_1` as an example for a simple linear regression model (Figure 9.2). The term 'simple linear regression' refers to linear regression models with only one predictor variable. Models with multiple predictors are called 'multiple linear regression models' and we will treat those later. The example dataset consists of leaf respiration rates (response variable `leafresp`) measured at various temperatures (explanatory variable `temp`). Before we start modelling, we should always do a sanity check by examining the structure (function `str`) of our data (dimensions of the dataset and type of variables) and a data summary (`summary`) to ensure that we are working with clean quality data (i.e. you should be able to flag out-of-range values or suspicious data points that might need looking at).

```
## Load required packages
> library(MASS)
> library(dplyr)

## Read in the leaf respiration data
> lr <- read.csv("leaf_respiration_1.csv")

## Check the structure of the data
> str(lr)
'data.frame':    100 obs. of  2 variables:
 $ leafresp: num  3.95 8.04 7.46 4.7 6.37 6.16 5.98 3.82 6.67 2.89 ...
 $ temp    : num  16.2 18.2 19.5 15 17.7 ...
```

Both `str` and `summary` are very useful functions, apply them to any object to learn more about it!

The `str` output tells us that we are dealing with a dataset consisting of 100 observations (rows) and 2 numeric variables (columns).

```
## Check the summary of the data
> summary(lr)
```

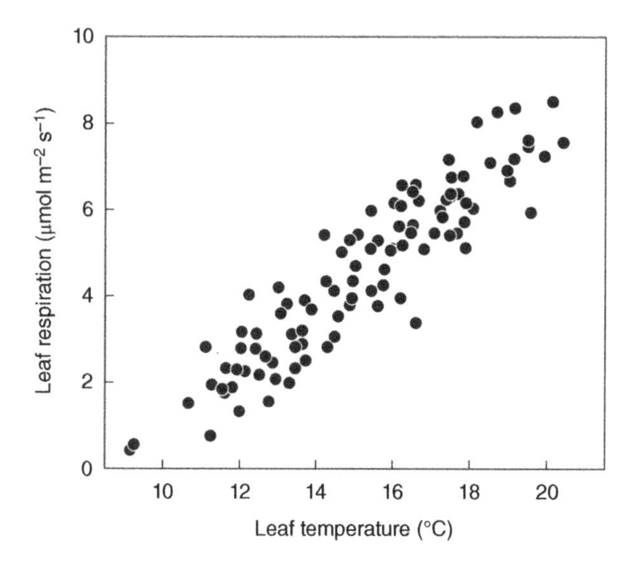

Figure 9.2 Leaf respiration as a function of leaf temperature. The relationship shows a clear linear trend, so choosing linear regression as our tool to analyse the dataset seems justified.

```
   leafresp          temp
Min.   :0.430   Min.   : 9.16
1st Qu.:2.873   1st Qu.:13.06
Median :4.660   Median :15.43
Mean   :4.561   Mean   :15.21
3rd Qu.:6.107   3rd Qu.:17.41
Max.   :8.510   Max.   :20.41
```

The `summary` output shows reasonable ranges for both variables (e.g. leaf respiration values are all positive and in line with what is reported in the literature).

After our initial check, we should plot the data to get a visual impression of the relationship. Note that the code below will produce a simpler default plot, rather than the one shown in Figure 9.2, e.g. with no units, etc., see Chapter 6 to learn how to produce custom graphs!

```
> plot(leafresp ~ temp, data = lr)
```

The scatterplot shows a clear linear relationship between leaf respiration and leaf temperature, making a linear regression model a reasonable choice. We can use the `lm` function (short for linear model) to perform a linear regression with these data and obtain the model summary like this:

```
> m1 <- lm(leafresp ~ temp, data = lr)
> summary(m1)

Call:
lm(formula = leafresp ~ temp, data = lr)

Residuals:
     Min       1Q   Median       3Q      Max
-2.13035 -0.54320  0.00733  0.45870  1.54416
```

```
Coefficients:
            Estimate Std. Error t value Pr(>|t|)
(Intercept) -5.84233    0.45642  -12.80   <2e-16 ***
temp         0.68390    0.02957   23.13   <2e-16 ***
–
Signif. codes:  0 '***' 0.001 '**' 0.01 '*' 0.05 '.' 0.1 ' ' 1

Residual standard error: 0.7729 on 98 degrees of freedom
Multiple R-squared:  0.8452,	Adjusted R-squared:  0.8436
F-statistic: 534.9 on 1 and 98 DF,  p-value: < 2.2e-16
```

The model summary may seem a little overwhelming when you see it for the first time. Therefore, we devote the following section to picking apart the summary output.

9.1.1 Making Sense of the `summary` Output for Regression Models Fitted with `lm`

Let us perform an 'autopsy' on the `summary` output of a simple regression model to build a firm grasp of all the components. Note that the output of `lm` without using `summary` only provides you with the estimated parameters, namely the intercept and the slope. For the keen reader, we also provide R code on how the individual components of an `lm` output can be computed.

- `Call` – Displays your model equation (the tilde symbol ~ reads 'as a function of').

```
Call:
lm(formula = leafresp ~ temp, data = lr)
```

- `Residuals` – A five number summary of the model residuals (minimum, first quartile – 1Q, median, third quartile – 3Q, maximum). These are the numbers that commonly form a boxplot and hence give you a quick idea about the spread of your residuals, which should be roughly symmetrical around the median. The median itself should be close to zero.

```
Residuals:
     Min       1Q   Median       3Q      Max
-2.13035 -0.54320  0.00733  0.45870  1.54416
```

We can use the `quantile` function to recreate this output like this:

```
quantile(resid(m1))
```

- `Coefficients` – A table containing the estimates of the model coefficients along with their standard errors and the associated *t*- and *P*-values.

Interpretation: The `intercept` (–5.84) gives the estimated value of the response variable when the explanatory variable is zero (the intercept may not always have a direct biological interpretation like in our case, but do not worry about that for now). The coefficient `temp` indicates the slope of the relationship between leaf respiration and temperature, i.e. the model predicts that leaf respiration increases by 0.68 $\mu mol\ m^{-2}\ s^{-1}$ for every degree Celsius. The standard errors are a measure of uncertainty around the coefficient estimates.

```
Coefficients:
             Estimate Std. Error t value Pr(>|t|)
(Intercept) -5.84233    0.45642  -12.80   <2e-16 ***
temp         0.68390    0.02957   23.13   <2e-16 ***
```

So, what is the story with these t-values? They are simply the coefficient estimates divided by their standard errors – a signal-to-noise ratio, so to speak, and this ratio follows a t-distribution. Let us exemplify this with the slope: 0.68390/0.02957 = 23.13 (you can find the unrounded results in `summary(m1)$coef` or `coef(summary(m1))`).

A coefficient estimate divided by its standard error gives the t-statistic!

Finally, the P-values (P stands for probability) are linked to the t-statistics and tell us whether the coefficients are significantly different from zero. This may sound strange, but it is actually quite easy to understand: a slope of zero represents a horizontal line, i.e. no change in the response variable with changes in the explanatory variable.

More formally, one would say the P-value is the probability of achieving an *absolute* value of the t-statistic as large or larger than the one at hand *if the null hypothesis were true.* The null hypothesis assumes that there is no effect, which means a parameter value of zero. So, a small P-value (conventionally <0.05) suggests that the coefficient estimate is statistically significant, meaning it is significantly different from zero.

In our example, we can derive the P-values from the t-distribution (using the cumulative distribution function `pt`) by hand like this:

```
2 * pt(q = abs(coef(summary(m1))[, 3]), df = df.residual(m1),
lower.tail = F)
```

where `pt` returns the probability for a certain value following a t-distribution, `abs(coef(summary(m1))[, 3])` extracts the two t-values from the summary output (`abs` ensures negative values turn positive), and `df.residual` returns the residual degrees of freedom (number of observations – number of model coefficients).

You may now wonder why we need to multiply the probabilities by 2 and why we specify `lower.tail = F`?

Because the associated t-tests are two-tailed, we have to multiply by 2. As t-values can be positive or negative, we simply use their absolute value (`abs` function) and then specify `lower.tail = F` to make sure the function provides the probability to the right (or upper tail) of the critical t-value, see Chapter 3.

```
## Probability of obtaining a t-value as small as this or smaller
## (more negative) => P-value for the lower tail
> pt(q = -12.80032, df = 98)
[1] 6.152912e-23

## Probability of obtaining a t-value as large as this or larger =>
## P-value for the upper tail
> pt(q = 12.80032, df = 98, lower.tail = F)
[1] 6.152912e-23

## Add both P-values to get the correct P-value for a two-tailed test
> pt(q = -12.80032, df = 98) + pt(q = 12.80032, df = 98, lower.tail = F)
[1] 1.230582e-22

## Boils down to...
> 2 * pt(q = 12.80032, df = df.residual(m1), lower.tail = F)
[1] 1.230582e-22
```

- `Signif. codes` – Commonly used shorthand notation using asterisks to indicate various significance levels. One asterisk indicates a P-value greater than 0.01 and smaller than 0.05, two asterisks signify $0.001 > P < 0.01$, and three asterisks mean highly significant with $P < 0.001$.

Reporting exact P-values (up to three digits) is always preferable to stating that P is larger or smaller than a threshold value, unless P is smaller than 0.001.

```
Signif.codes: 0 '***' 0.001 '**' 0.01 '*' 0.05 '.' 0.1 ' ' 1
```

- `Residual standard error` – This is the standard error of the residuals, which is called sigma (σ) and serves as an estimate of the true model error. The residual standard error is a measure of quality of a linear regression fit and represents the average deviation of the observed values (response values) from the regression line. It has the same unit as the response variable.

```
Residual standard error: 0.7729 on 98 degrees of freedom
```

The residual standard error can be calculated using the formula provided in Chapter 2, so in R:

$$\sigma = \sqrt{\frac{\sum_{i=1}^{n}(y_i - \hat{y}_i)^2}{n - p}}$$

where y denotes the response variable, $\hat{y}$ indicates the associated fitted (predicted) value, n is the number of observations, and p is the number of model parameters (model coefficients), with $n - p$ signifying the residual degrees of freedom.

We can check this by typing:

```
sqrt(sum(resid(m1)^2)/df.residual(m1))
```

- `Multiple R-squared` – Coefficient of determination, a value in the 0–1 range that indicates how well the model fits the data. Generally, the higher the R^2, the better the model fits the data. It can be thought of as the proportion of the variation in the response variable that is explained by the model. Here, the model explains c. 85% of the variation.

```
Multiple R-squared:  0.8452
```

The R^2 value is calculated as:

$$R^2 = 1 - \frac{\text{Residual sum of squares (RSS)}}{\text{Total sum of squares (TSS)}}$$

In our example, we can calculate the R^2 by hand like this:

```
1 - (sum(resid(m1)^2)/sum((lr$leafresp - mean(lr$leafresp))^2))
```

- `Adjusted R-squared` – R^2 value adjusted for model complexity, i.e. the number of model parameters (coefficients). This is a necessary adjustment as the R^2 will get larger by pure chance, the more parameters we include in the model regardless of their significance. This is because more parameters give the model more 'wiggle room' and hence yield a closer fit that explains more variation.

```
Adjusted R-squared:  0.8436
```

The adjusted R^2 can be calculated as follows:

$$R^2_{\text{adj.}} = 1 - (1 - R^2)\,\frac{n - 1}{n - p}$$

where n is the number of observations and p gives the number of model parameters (here the intercept and slope).

We can calculate the adjusted R^2 ourselves like this:

```
1 - (1 - summary(m1)$r.squared) * (100 - 1)/(100 - 2)
```

Always indicate the adjusted R^2 rather than the multiple R^2.

- `F-statistic` – A measure of the overall significance of the regression model. It is the ratio of the variance explained by the model parameters (ESS) to the residual (unexplained) variance (RSS) standardised by their respective degrees of freedom (see Figure 9.1).

```
F-statistic: 534.9 on 1 and 98 DF
```

The formula for the F-statistic is given by:

$$F = \frac{(\text{ESS}/k)}{(\text{RSS}/(n - k - 1))}$$

where ESS is the explained sum of squares and k represents the associated degrees of freedom given by the number of explanatory variables (just one in a simple regression model). RSS indicates the residual sum of squares with $n - k - 1$ degrees of freedom, where n is the number of observations.

In R, we can obtain the F-value by hand like this:

```
(sum((fitted(m1) - mean(lr$leafresp))^2)/1)/
(sum(resid(m1)^2)/(100-1-1))
```

To aid understanding, we can also consult the analysis of variance table of our regression model using `anova(m1)`, which gives the results of testing the null hypothesis that all model coefficients are equal to zero (have no effect). The output contains the explained and residual sum of squares and their degrees of freedom, and the resulting F-statistic along with the associated P-value. To avoid confusion at this stage, we leave it to the interested reader to examine the `anova(m1)` output.

- `P-value` – The P-value associated with the above F-value. The P-value gives the probability of achieving an F-value as large as this or larger under the null hypothesis (no effect). This P-value gives an overall significance of the model and is commonly reported alongside the R^2 value to give the reader reassurance that the amount of explained variation is actually statistically significant.

```
p-value: < 2.2e-16
```

It is easy to calculate the P-value by hand using the distribution function of the F-distribution (with `lower.tail = F` because we are interested in the probability of obtaining equal or larger F-values only by chance; see Sections 3.4.1 and 3.6):

```
pf(534.9, df1 = 1, df2 = 98, lower.tail = F)
```

9.1.2 Model Diagnostics

We need to check the underlying model assumptions to ensure the reliability and trustworthiness of the model. The most basic assumption of a linear model is the presence of a linear relationship between the response and the predictor variable(s), which can be checked prior to model fitting using bivariate plots with the response on the y-axis and the predictor on the

x-axis. However, most assumptions require us to look at the unexplained variation (residual variation) that cannot be explained by the model. This means that these assumptions can only be assessed *after* model fitting. Linear regression assumes:

1. Linearity, i.e. the relationship between the response variable and the explanatory variable is assumed to be linear. This means that a given change in the explanatory variable is associated with a proportional change in the response variable.
2. Variance homogeneity (*homoscedasticity*), i.e. constant variance across the range of the continuous predictor(s).
3. Normally distributed errors in the y-space (recall that the model errors are estimated by the residuals), but no errors in the x-space (i.e. the predictor is measured without error).
4. Independence (as in independent observations and thus independent residuals, e.g. repeated measures on the same subject are not independent, eggs sampled from the same nest are not independent, strictly speaking, trees growing in the same forest patch are not independent either).
5. No outliers or influential points as these may result in biased parameter estimates. Influential data points can be outliers, but not all outliers are necessarily influential. The impact of an influential data point depends on its leverage (impact on parameter estimation) and how much it affects the model fit.

The model assumption of independence cannot be tested post hoc, this is something that you have to consider while planning your experiment!

Model diagnostics commonly focus on the variance homogeneity and normality criteria, which are best investigated graphically using so-called *residual plots*. But why are we looking at the residuals, and not the original data?

We will clarify this in an intuitive way. Statistical models rarely encompass the entire population but are usually based on a sample. For example, if we are interested in the relationship between stem diameter and height of a tree species, we would take sample measurements of several individuals, but we would never measure *all* existing individuals of this species (the entire population). Therefore, we will never know the *true* coefficients (intercept and slope). All we can come up with is a best estimate based on our sample, which is the reason why the model coefficients are also referred to as *estimates*. Just like the model coefficients are estimates of the true population coefficients, the model residuals are estimates of the true error (difference between observations and model predictions). The concept of using the residuals to check the underlying model assumptions may seem strange at first but it makes good sense because:

Assessing diagnostic plots needs practice, and there is no firm 'threshold' past which an assumption is violated.

In a model that describes your data well, the predictor(s) explain(s) most of the variation seen in the response, and thus the remaining, unexplained variation captured by the residuals should only represent the inherent randomness of any real-world phenomenon. But what do random errors (residuals) look like? The short answer is: Like a starry sky (Figure 9.3a)!

The residuals should not be systematically high or low but should be centred around zero with fairly constant spread along the range of fitted values. Such a pattern indicates variance homogeneity and suggests that the model is adequate for all fitted values (model predictions). These assumptions can be elegantly assessed by plotting the residuals vs. the fitted values (Figure 9.3). Unfortunately, variance homogeneity is quite uncommon. Frequently, we encounter fan-shaped patterns or some sort of nonlinear trends (Figure 9.3b,c). This is where we need to start playing detectives to find out where these patterns come from, using our subject matter expertise. Is the pattern caused by the predictor variable? (Plot the residuals against the predictor variable to find out). Are we perhaps missing important predictors or interactions among them? (Include more predictors and/or interactions in the

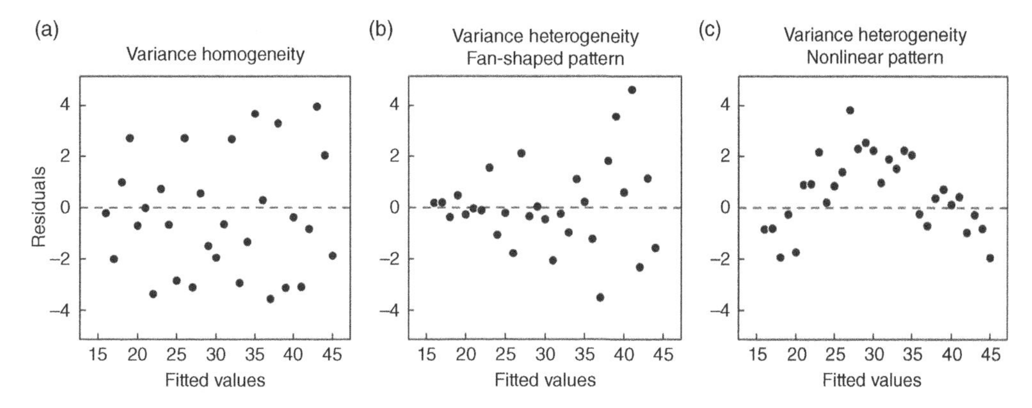

Figure 9.3 Plots of residuals vs. fitted values. (a) A residual pattern indicating variance homogeneity. (b) Variance heterogeneity: a fan-shaped residual pattern showing increasing variance with increasing fitted values. Often, this pattern can already be discerned in a plot of the raw data (response vs. predictor) if the response values scatter more strongly with increasing predictor values. (c) Variance heterogeneity: a hump-shaped residual pattern suggesting a nonlinear trend in the data, which may be indicative of an inappropriate model fit.

model and reassess the diagnostic plots, see Section 9.2 and Chapter 11.) Do we have correlated errors due to repeated measures, which would violate the model assumption of independent observations? (Reflect carefully on the experimental design and implementation. Look for identical units, subjects, or other identifiers appearing in multiple rows of your dataset.)

By now you should start to realise why, contrary to intuition, we need to fit a linear model first in order to check the underlying assumptions and determine whether it is appropriate to apply the model in the first place. Inspecting histograms of the response variable prior to model fitting, and perhaps even running a formal test, to determine whether the response variable follows a normal distribution are redundant because remember: the normality assumption of linear models applies to the errors (residuals), NOT the response variable. Checking the response variable for normality can only give you a weak indication as to whether the residuals might be normally distributed. In fact, sometimes you may find histograms of response variables that suggest anything but a normal distribution while the model residuals show a normality pattern that could have come straight from a textbook.

Counter-intuitively, we must first fit a model to check if the underlying assumptions are met (as model diagnostics require residuals).

Since the model residuals simply represent the leftover variation of the response variable after model fitting, we can readily calculate them by subtracting the fitted values from the observed (measured) response values (Figure 9.1).

Violations of variance homogeneity commonly introduce a greater bias than deviations from normality (Zuur et al. 2009). In practice, we therefore get away with a great deal of non-normality. However, variance heterogeneity (*heteroscedasticity*) often results in seriously biased (often inflated) standard errors and thus biased test statistics and *P*-values. R offers a range of diagnostic plots that can be produced by passing a model object to the `plot` function (Figure 9.4).

```
## Model diagnostic plots
> par(mfrow = c(2, 2))
> plot(m1)
```

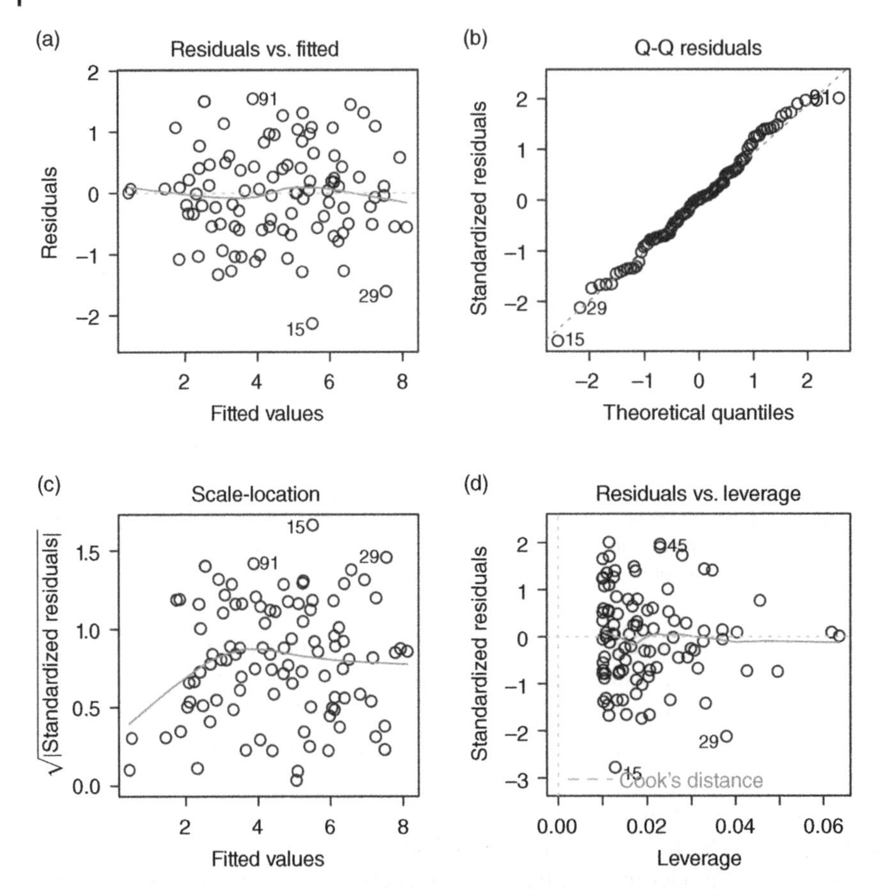

Figure 9.4 Model diagnostic plots. (a) **Residuals vs. fitted (predicted) values** – for assessing the **variance homogeneity** (homoscedasticity) criterion (a random pattern suggests variance homogeneity, variance heterogeneity is commonly detected by fan-shaped or undulating patterns). (b) **Quantile-quantile plot** contrasting the quantiles of the residuals with those of a standard normal distribution to assess the normality criterion (if the residual distribution approximates a normal distribution, then the points fall on one line). (c) **Scale-location plot** similar to the first plot but uses the square root of the standardised residuals instead to improve pattern recognition. (d) **Standardised residuals vs. leverage plot** – The standardised residuals (raw residuals divided by their standard deviation) indicate how many standard deviations an observation varies from its prediction and helps with outlier detection. Leverage is a measure of how strongly an individual data point influences the estimated coefficients of the model. High or low data points far to the right are of particular interest. Data points labelled with their observation number are worthwhile inspecting. A dashed grey contour line indicates the Cook's distance, another measure of influence that incorporates leverage. However, this line is not displayed here because our data are below the critical Cook's distance threshold.

In the case of an `lm` object, four diagnostic plots are created by default (Figure 9.4):

- Top left: ***Plot of the model residuals vs. fitted values*** – Under the constant variance assumption, we would expect to see a relatively constant spread of the residuals over the range of the fitted values (predicted response values). These plots may also help uncover any nonlinear trends in the residuals, which would suggest a nonlinear relationship between the predictor and the response variable that was not captured by the model. Ideally, this plot shows no discernible trend (i.e. you should see a starry sky pattern).

- Top right: ***Quantile-quantile plot of the standardised residuals (Q-Q plot)*** – When we plot the quantiles of our standardised model residuals (raw residuals divided by their standard deviation) against the theoretical quantiles of a standard normal distribution, then we would expect the points in a normal Q-Q plot to fall on a straight line. Deviations from a straight line indicate non-normality.
- Bottom left: **Scale-location plot** – This is a variation of the top left plot using the square root of the absolute values of the standardised residuals (residuals divided by their standard deviation) instead of the raw residuals. Using this type of residuals makes it potentially easier to pick up violations of the homoscedasticity assumption.
- Bottom right: ***Standardised residuals vs. leverage plot*** – With this diagnostic tool, we assess the influence of individual data points (observations) on the fit of a statistical model. In this plot, overall patterns are not important. Here we look for positive and negative extreme values on the y-axis and values far to the right of the x-axis. This plot helps identify outliers and/or influential observations. The values on the y-axis indicate how many standard deviations an observed value deviates from the fitted value and observations beyond a certain threshold (e.g. ± 3 standard deviations) may be considered outliers. The x-axis represents the leverage, which is a measure of how strongly an individual data point influences the estimated coefficients of the model. Points toward the right of the plot with high leverage indicate highly influential observations. To clarify terms: ***Outliers*** are observations whose standardised residuals deviate strongly from zero (extreme values) and may or may not have a large impact on the model fit, i.e. not all outliers are influential. ***Influential points*** strongly affect the coefficient estimates and the overall fit of the model due to high leverage. In conclusion, not all influential points are outliers, and *vice versa*. In the presence of very large values (on the y- or x-axis), another measure of influence, the Cook's distance D is displayed as a grey dashed contour line and observations outside of it are considered influential. Cook's distance is defined as the normalised change in fitted values associated with the omission of an observation during the modelling process, *i.e.* the model is iteratively refitted and each time a different observation is left out.

In an extreme case, a single data point can reverse a regression line, here, the black one is the culprit!

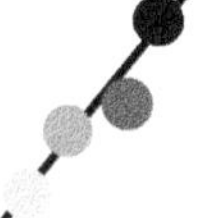

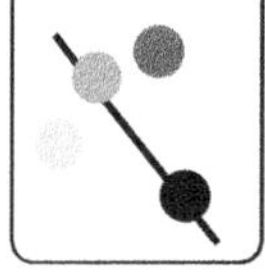

In summary, the diagnostic plots do not indicate gross violations from homoscedasticity or normality and have not flagged any outliers or influential data points that we need to worry about (all within the critical Cook's distance). This implies that the model is trustworthy and provides a reasonable representation of the relationship between our response and the explanatory variable.

A modern version of the model diagnostic plots based on *ggplot2* graphics can be found in package *performance* (function `check_model`, Lüdecke et al. (2021)).

9.1.3 Model Predictions and Visualisation

Once you have successfully fitted a linear model, verified the diagnostic plots, and looked at the significance of the parameters, you can visualise the results and use them to make predictions. For example, you can use the `predict` function to obtain model predictions for any given x-value:

To predict a single value, you can simply multiply your x-value by the slope and add the intercept!

```
## Using the linear model to predict leaf respiration from leaf temperature
> predict(m1, newdata = data.frame(temp = 15))
       1
4.416114
```

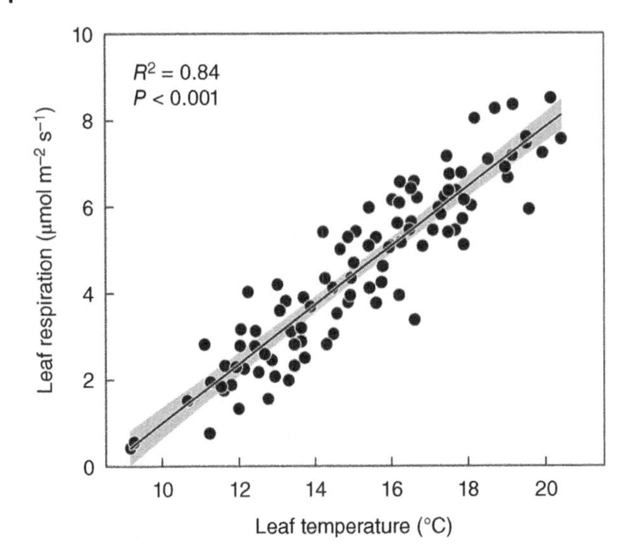

Figure 9.5 Leaf respiration as a function of leaf temperature. The black line represents the linear regression fit and the surrounding grey area indicates the 95% confidence interval.

It is important to note that the name of the predictor variable in the data frame supplied to the `newdata` argument must be identical to that in the model, otherwise the function will not work. To predict over a range of values, for example the original range of leaf temperatures, we can pre-define a data frame and add the model fit along with a 95 % confidence interval like this:

```
## Create a new data frame to predict from, providing a fine grid of
predictor values spanning the original range of leaf temperatures
> newdat <- data.frame(temp = seq(min(lr$temp, na.rm = T), max(lr$temp,
na.rm = T), length.out = 200))

## Get model predictions and associated 95 % confidence intervals
> preds <- predict(m1, newdata = newdat, interval = "confidence", level =
0.95)
> newdat <- bind_cols(newdat, preds) # combine all into one dataframe

## Sanity check
> head(newdat, n = 3)
      temp       fit        lwr       upr
1 9.160000 0.4221595 0.03531195 0.8090071
2 9.216533 0.4608220 0.07701775 0.8446262
3 9.273065 0.4994845 0.11871897 0.8802500
```

To visualise the model, we simply add the regression line along with the R^2 value and the P-value for the overall significance to the scatter plot (Figure 9.5):

```
## Scatter plot with regression line, 95 % confidence interval (CI) and
statistical annotation
> plot(leafresp ~ temp, data = lr)
> with(newdat, polygon(x = c(temp, rev(temp)), y = c(lwr, rev(upr)), col =
"grey75", border = "grey75")) # polygon indicating the 95 % CI
> points(leafresp ~ temp, data = lr, pch = 21, bg = "black")
```

```
> # abline(m1) # adds the regression line over the whole plotting region
> lines(fit ~ temp, data = newdat) # regression line in the predictor range
> text(x = 9.5, y = 8.3, label = expression(paste(italic("R")^2, " = 0.84")),
adj = c(0, 0.5))
> text(x = 9.5, y = 7.7, label = expression(paste(italic("P"), " < 0.001")),
adj = c(0, 0.5))
```

9.1.4 What to Write in a Report or Paper?

Correctly talking about the method of linear regression in a methods section, but also correctly reporting the results of a linear regression analysis are important. While there is not one single right way of wording this, we provide you with a template that covers all the important information you should include.

9.1.4.1 Material and Methods

All statistical computations were performed using the R software environment (R Core Team 2022).

We used a simple linear regression model to analyse the relationship between leaf respiration and leaf temperature. The variance homogeneity and normality assumptions of the model errors were assessed using diagnostic plots (normalised residuals vs. fitted values and quantile-quantile plots). The residual plots did not indicate any gross violations of these model assumptions. Based on the Cook's distance (D) and its commonly used threshold of $D < 0.5$, we did not detect any potentially influential data points.

Remember – R is free, so citing R and R packages correctly is important, it gives the authors credit! See help file for `citation`.

9.1.4.2 Results

Leaf respiration increased significantly in response to rising leaf temperature with a slope of 0.68 ($F_{1,98} = 534.9$, $P < 0.001$). Leaf temperature explained about 84% of the variation in leaf respiration.

9.1.5 Dealing with Variance Heterogeneity

Biological or environmental data almost always exhibits a certain amount of variance heterogeneity, which needs to be addressed, otherwise this unaccounted-for variation will result in inflated standard errors of the model parameters (and thus a smaller test statistic, increasing the likelihood of a type II error). In some cases, missing explanatory variables or missing interactions may be the culprits and including them in the model may improve the variance pattern (Section 9.2 and see Chapter 11). Often, however, including extra variables is not possible or simply does not help. In those cases, *variance modelling* using generalised least squares (GLS) models that allow the incorporation of variance structures is the best solution (switching from the `lm` function to the `gls` function in package *nlme*, Pinheiro et al. 2023; Pinheiro and Bates 2000; see Chapter 12 for a worked example). Such GLS models relax the assumption that errors must have equal variances and/or must be independent. These models allow the incorporation of so-called variance functions to deal with variance heterogeneity (heteroscedasticity) and/or correlation functions to handle correlated errors, which we commonly encounter with repeated measures data (temporal autocorrelation). In previous times, when nifty options like variance modelling using GLS regression were not widely available, people commonly used to apply a *variance stabilising data transformation*, usually a *logarithmic* or *square root transformation* of the response variable (and we still see this old-school approach in a lot of publications today). This is quite a brute-force approach as the transformation distorts the original relationship between response and predictor and can make model interpretation difficult. With modern modelling tools at our fingertips, data transformation should be considered a last resort. If you use it, make sure both the response variable and the predictor(s) are transformed to ensure that the original relationship is maintained on the transformed scale (transform-both-sides approach). The choice of transformation should ideally be guided by the Box-Cox approach (`boxcox` function, R package MASS, Venables and Ripley 2002), which optimises the so-called transformation parameter (λ) to achieve the best stabilisation of variance and thus helps identify the most suitable data transformation (Section 8.3).

Transformations can also be useful to better visualise data, see Chapter 8.

For the sake of completeness, we demonstrate a data transformation procedure but strongly encourage you to familiarise yourself with the more sophisticated and powerful variance modelling approach using GLS models (R package *nlme*). We will use a slightly modified version of our leaf respiration dataset displaying increasing spread with rising leaf temperature (Figure 9.6a). The diagnostic plots of a simple linear regression with these data show that the larger variation at higher leaf temperatures is reflected in the residuals in the form of a fan-shaped pattern (Figure 9.7a). This is a strong indication of variance heterogeneity, which can also be seen in Figure 9.7c as an increase in residual values with increasing fitted values. We will try to eliminate the variance heterogeneity using a data transformation.

```
## Load data and plot it
> lr2 <- read.csv("leaf_respiration_2.csv")
> plot(leafresp2 ~ temp, data = lr2)

## Run linear model
> m2 <- lm(leafresp2 ~ temp, data = lr2)

## Diagnostic plots
> par(mfrow = c(2, 2))
> plot(m2)
```

But which data transformation should we apply? The Box-Cox method indicates that the optimal λ is 0.33 (i.e. 1/3), meaning that all observations should be raised to the power of 1/3, which translates into a cube root transformation (Figure 9.6b). The 95% confidence interval of λ includes 0.5, suggesting that the square root transformation could also be an option. We will add two new columns to our dataset with the cube root transformed response and predictor variables. The resulting plot looks much more 'well-behaved' in terms of variance relative to the untransformed data (compare Figure 9.6a and c). Now we run another linear regression model using the cube root transformed data and check whether the transformation has resolved the variance heterogeneity issue. We use the `mutate` function (R package *dplyr*) to add the cube root transformed variables to the dataset. The `mutate` function provides more flexibility than base R functions, which will become more relevant with increasing coding proficiency and more challenging data manipulation tasks. However, if `mutate` appears alien to you, then you can add a variable the traditional way (see Section 1.2.1.3).

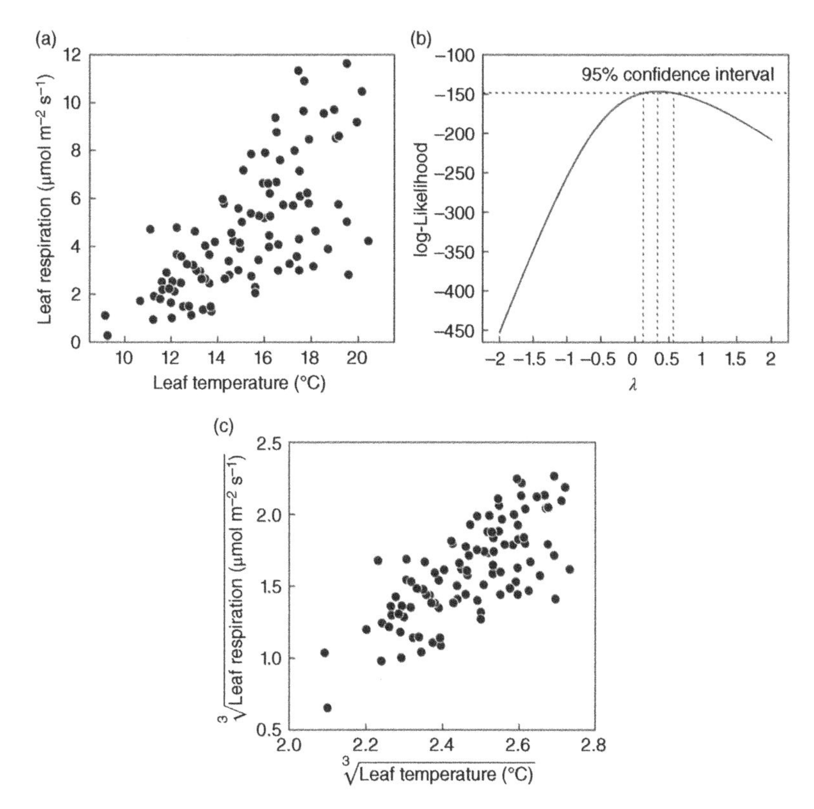

Figure 9.6 (a) Leaf respiration as a function of leaf temperature, original data. (b) The Box-Cox transformation (see Section 8.3 for details) shows that a value of the transformation parameter λ of 0.33 maximises the log-likelihood function, indicating that the best transformation to achieve variance homogeneity and normality is a cube root transformation (all observations are raised to the power of λ, i.e. $y^{1/3} = \sqrt[3]{y}$). The 95% confidence interval includes 0.5, suggesting that the square root transformation could be a viable alternative. (c) Original data after cube root transformation. Note the strongly increasing spread of the response variable with increasing leaf temperature in the original data (a) in contrast to the relatively constant variance across leaf temperatures after cube transformation of the data (c).

```
## Add cube root transformed variables to the dataset
> lr2 <- mutate(.data = lr2, leafresp2_cr = leafresp2^(1/3),
                             temp_cr = temp^(1/3))

> head(lr2, n = 3)
  leafresp2  temp leafresp2_cr  temp_cr
1  3.986123 16.20     1.585563 2.530298
2  4.644784 18.18     1.668483 2.629448
3 11.633331 19.51     2.265868 2.692066

> plot(leafresp2_cr ~ temp_cr, data = lr2)

## Simple linear regression using the cube root transformed data
> m3 <- lm(leafresp2_cr ~ temp_cr, data = lr2)

## Diagnostic plots
> par(mfrow = c(2, 2))
> plot(m3)
```

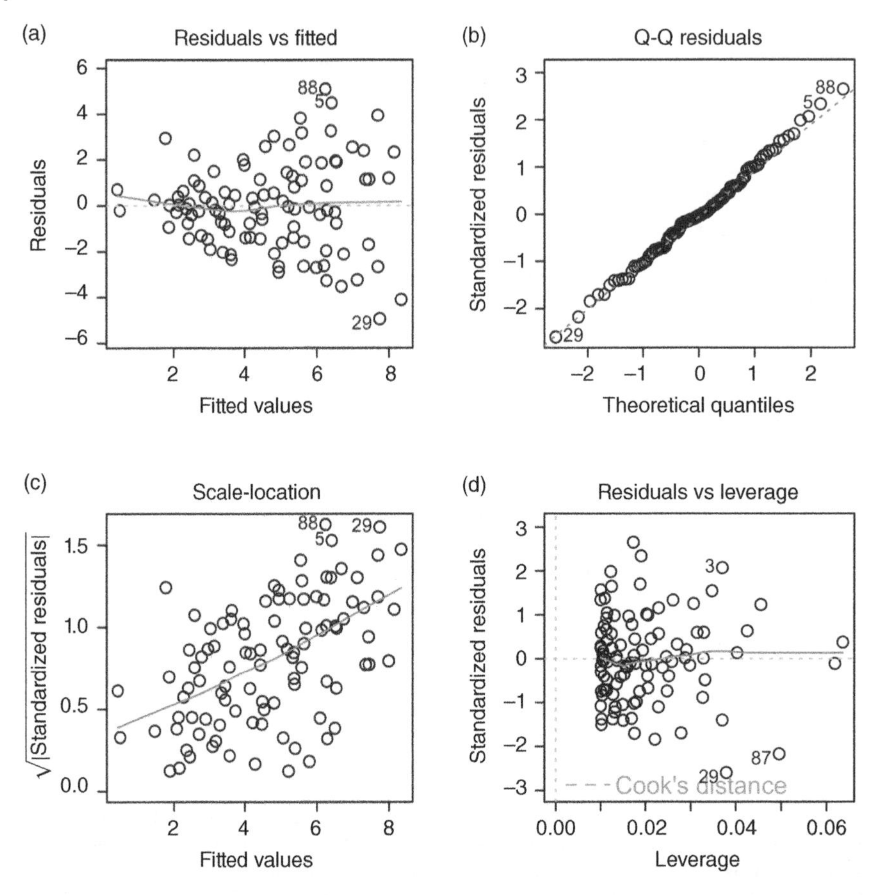

Figure 9.7 Diagnostic plots of the simple linear regression model m2, based on the more variable leaf respiration dataset. (a) The residuals vs. fitted values plot shows a strongly fan-shaped pattern indicating heteroscedasticity. (b) The larger variation in the leaf respiration data did not cause deviations from normality of the model errors. (c) The scale-location plot confirms the increasing spread of the model errors with larger fitted values. (d) The residuals vs. leverage plot highlights a couple of observations, but these lie well within the critical Cook's contour line (dashed grey line), which is not visible because it is outside the plotting region.

Indeed, the cube root transformation has largely eliminated the fan-shaped pattern in the residuals vs. fitted values plot and did not change the distribution of the residuals much from normality as evidenced by the Q-Q plot (Figure 9.8). With no violations of the underlying assumptions, we can now safely use model m3 for inference and prediction.

9.2 Multiple Linear Regression

Up to this point, we have only dealt with simple linear regression, i.e. a model with a single continuous predictor. When we have two or more continuous predictors, we talk about multiple linear regression. We will use an extension of the leaf respiration data containing

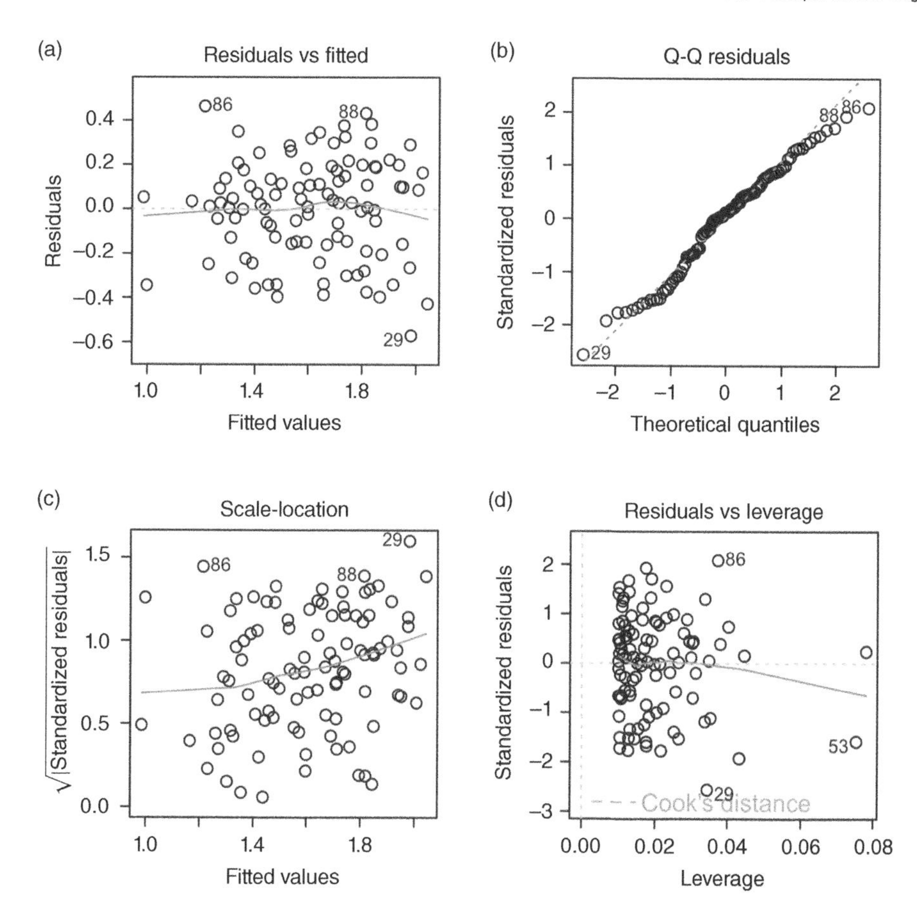

Figure 9.8 Diagnostic plots of the simple linear regression model m3 using cube root transformed data. The residuals vs. fitted values plots in (a) and (c) still show a slight pattern, but overall the relatively constant spread indicates no gross deviations from homoscedasticity. The Q-Q plot (b) and the residuals vs. leverage plot (d) do not indicate any issues in terms of normality or influential observations, respectively.

two additional explanatory variables: specific leaf area (SLA, m^2 kg^{-1}) and leaf nitrogen content (%). SLA (ratio of leaf area to leaf dry mass) is an important functional trait related to adaptations to the environment and to resource use strategies.

```
## Read in the data
> lr3 <- read.csv("leaf_respiration_3.csv")

## Check the structure
> str(lr3)
'data.frame':      100 obs. of  4 variables:
 $ leafresp: num  7.92 4.86 6.61 3.5 6.48 ...
 $ temp    : num  19 14.3 17.3 10.5 16.6 ...
 $ sla     : num  21.8 11.4 14.3 11.6 19.1 ...
 $ nitro   : num  2.39 1.83 2.05 1.69 2.2 ...


## Inspect the data summary
> summary(lr3)
```

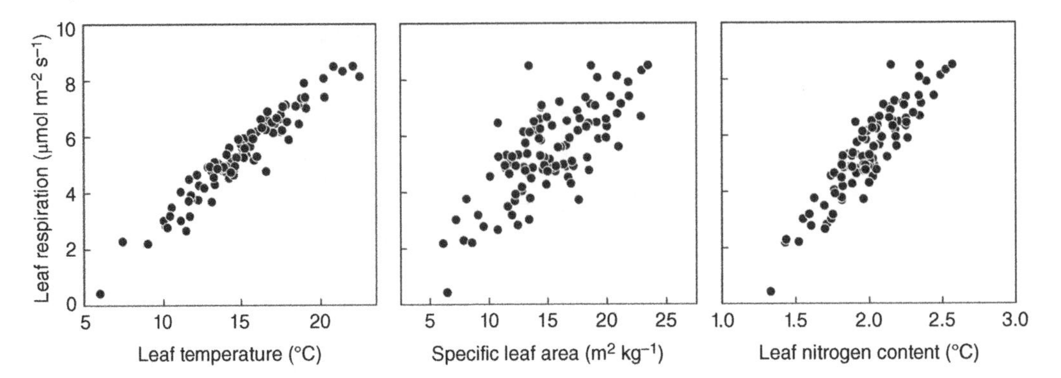

Figure 9.9 Linear relationships between leaf respiration and leaf temperature, specific leaf area, and leaf nitrogen content.

```
   leafresp            temp               sla                nitro
Min.   :0.430   Min.   : 5.987   Min.   : 6.102   Min.   :1.330
1st Qu.:4.651   1st Qu.:13.241   1st Qu.:12.736   1st Qu.:1.825
Median :5.309   Median :15.039   Median :14.651   Median :1.990
Mean   :5.416   Mean   :15.063   Mean   :15.071   Mean   :1.993
3rd Qu.:6.492   3rd Qu.:17.068   3rd Qu.:17.646   3rd Qu.:2.150
Max.   :8.510   Max.   :22.484   Max.   :23.477   Max.   :2.570
```

The SLA values are consistent with the figures cited in the literature and leaf nitrogen commonly varies between 1% and 3%, indicating no inconsistencies in our data. It is always a good idea to plot the data first to get a visual impression of the relationships. As we can see in Figure 9.9, leaf respiration shows linear relationships with all three explanatory variables, suggesting that a linear regression approach seems appropriate.

We will start with a basic linear regression considering only the main effects of the three predictors but no interactions between them. We call this an additive model, and the predictors are added using '+' symbols.

```
## Linear regression model
> m4 <- lm(leafresp ~ temp + sla + nitro, data = lr3)
> summary(m4)

Call:
lm(formula = leafresp ~ temp + sla + nitro, data = lr3)

Residuals:
     Min       1Q   Median       3Q      Max
-1.22746 -0.25208  0.06253  0.29470  0.69580

Coefficients:
            Estimate Std. Error t value Pr(>|t|)
(Intercept)   32.861    119.840   0.274    0.785
temp           1.952      5.190   0.376    0.708
sla            1.545      5.189   0.298    0.767
nitro        -40.210    138.600  -0.290    0.772

Residual standard error: 0.405 on 96 degrees of freedom
Multiple R-squared:  0.9349,     Adjusted R-squared:  0.9328
F-statistic: 459.2 on 3 and 96 DF,  p-value: < 2.2e-16
```

Astonishingly, the summary output shows non-significant *P*-values for the slopes of all three predictors implying that none of the predictors has a significant effect on leaf respiration. This is a very surprising outcome, given the rather clear linear relationships (Figure 9.9). What is very striking are the very large standard errors, which are indicative of a condition called *multicollinearity*.

9.2.1 Multicollinearity in Multiple Regression Models

A frequent problem with multiple regression models is the phenomenon of multicollinearity. Multicollinearity occurs when two or more of the continuous predictor variables are highly correlated, i.e. one predictor variable can be linearly predicted from the others with high accuracy. Often, multicollinearity produces unstable parameter estimates and increases the variance and thus the standard errors of the coefficients (inflated standard errors) resulting in flawed *P*-values and poor overall model performance. In practice, this often happens when one or more of the predictors in a multiple regression model provide similar or complementary information. For example, 'relative humidity' and 'temperature' are highly correlated and will hence give rise to multicollinearity issues in a regression model. One cannot always tell from the model summary whether collinearity issues are present but there are two main ways of **detecting multicollinearity**:

1. Graphically, using scatterplot matrices (e.g. a pairs plot, see Chapter 6), where all explanatory variables are plotted against each other (this is a rather informal approach).
2. Using **variance inflation factors (VIFs)** (`vif` function in R package *car*, Fox and Weisberg 2019). VIF values larger than 10 suggest strong multicollinearity caused by the respective variable. VIF values between 5 and 10 indicate a moderate level of multicollinearity. VIF values < 5 are deemed acceptable, suggesting a relatively low degree of multicollinearity.
3. Using a hybrid approach with a compound visualisation of a τ ('tau') statistic that quantifies the extent of collinearity in the dataset, and a related multicollinearity index ranking the predictors by their contribution to collinearity (`mcvis` function in R package *mcvis*, Lin et al. 2020).

These tools can help you decide which explanatory variables to retain in the model and which ones to drop to eliminate multicollinearity issues. Here are all tools in action, starting with a pairs plot:

```
## Pairs plot
> library(ggplot2)# load ggplot2 package
> install.packages("GGally")
> library(GGally) # load GGally package
> ggpairs(lr3)
```

The pairs plot clearly shows that the predictors are not only linearly related with the response variable but also with each other, which is a strong sign of multicollinearity (Figure 9.10). The bivariate predictor relationships containing leaf nitrogen display the strongest correlations, which points to this predictor as the main cause of multicollinearity (but correlation alone is not a robust indicator of multicollinearity!).

For different types of pairs plots, refer to Chapter 6.

The `vif` function calculates VIFs for all predictors in a regression model. VIFs give a measure of how much the variance of the estimated regression coefficients is inflated compared

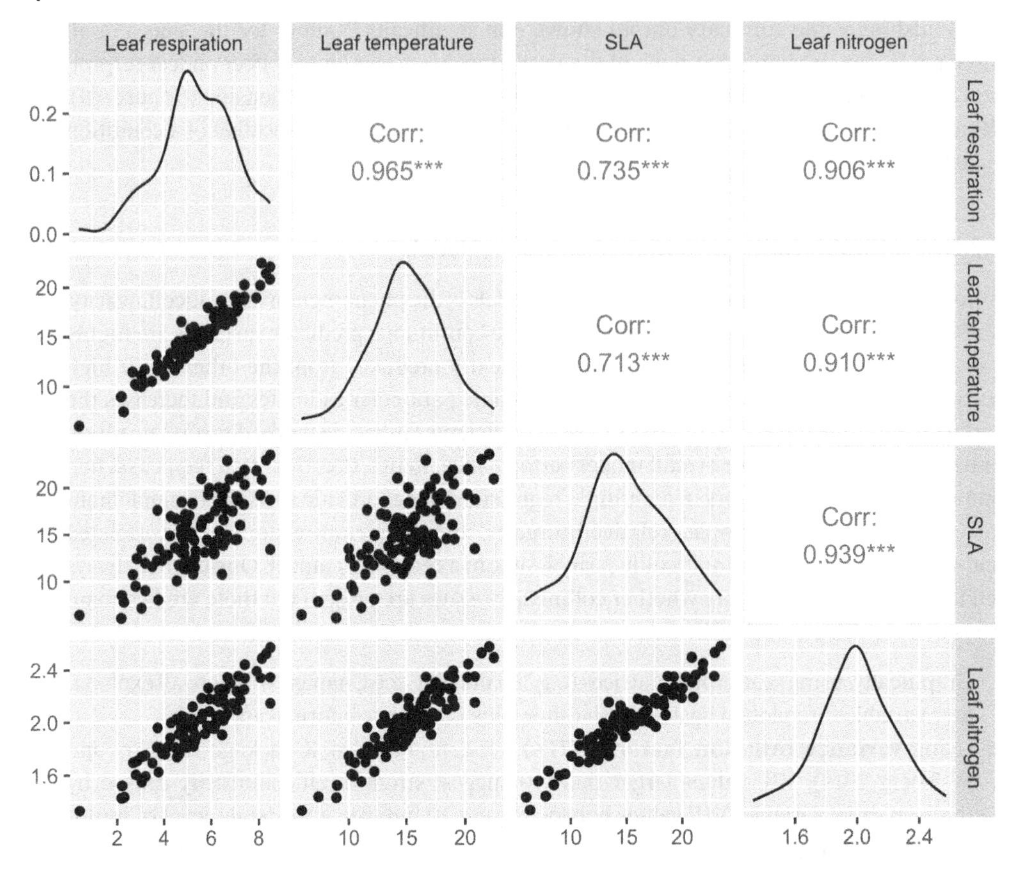

Figure 9.10 Pairs plot of the extended leaf respiration dataset. The scatterplot matrix shows kernel density plots for each variable in the diagonal plots to give an idea of their distribution. All bivariate relationships are displayed as scatterplots in the lower off-diagonal panels and the corresponding correlations in the upper off-diagonal panels. Reading example: the bottom scatterplot on the far left shows the leaf nitrogen–leaf respiration relationship and the uppermost panel on the far right shows their correlation of 0.906.

to when the predictor variables are not linearly related. So, a predictor with a VIF value of 10, for example, suggests that the variance around a model coefficient is 10 times larger than one would expect if there was no correlation with other predictors (in a model with no multicollinearity). Here are the VIFs related to our multiple regression model:

```
## Variance inflation factors (VIF)
> install.packages("car")
> library(car)
> vif(m4)
    temp       sla     nitro
159880.3 233352.0 668633.8
```

Similar to the pairs plot, the VIFs indicate that leaf nitrogen seems to be the main culprit and should be removed from the model.

A modern take on detecting multicollinearity comes with a new approach by Lin et al. (2020), leveraging the relationship between VIFs and specific properties of a matrix transformation (τ, pronounced 'tau', which is the inverse of the eigenvalues of the Gram matrix) to derive a multicollinearity index.

```
## Multicollinearity visualisation using a multicollinear-
ity index
## Install and load the mcvis package
> install.packages("mcvis")
> library(mcvis)
> mc <- mcvis(X = lr3[, c("temp", "sla", "nitro")]) # sup-
ply the predictors
> mc
     temp  sla nitro
tau3 0.12 0.23  0.67

> plot(mc)
```

The largest τ value (0.67) is associated with leaf nitrogen content and when visualised in conjunction with the multicollinearity indices (shown as lines of different grey shades and widths), leaf nitrogen is clearly identified as the main collinearity-causing variable (Figure 9.11). Indeed, removing leaf nitrogen from the model gets rid of the multicollinearity issue as indicated by the reasonable standard errors of the model coefficients and VIF values of the two remaining predictors <5.

```
## Simplified model without leaf nitrogen as predictor
> m5 <- lm(leafresp ~ temp + sla, data = lr3)
> summary(m5)

Call:
lm(formula = leafresp ~ temp + sla, data = lr3)

Residuals:
     Min       1Q   Median       3Q      Max
-1.23402 -0.25171  0.07657  0.29905  0.68146

Coefficients:
            Estimate Std. Error t value Pr(>|t|)
(Intercept) -1.90589    0.20127  -9.470 1.88e-15 ***
temp         0.44650    0.01842  24.238  < 2e-16 ***
sla          0.03955    0.01524   2.594    0.011 *
—
Signif. codes:  0 '***' 0.001 '**' 0.01 '*' 0.05 '.' 0.1 ' ' 1

Residual standard error: 0.4031 on 97 degrees of freedom
Multiple R-squared:  0.9348,       Adjusted R-squared:  0.9335
F-statistic: 695.4 on 2 and 97 DF,  p-value: < 2.2e-16
```

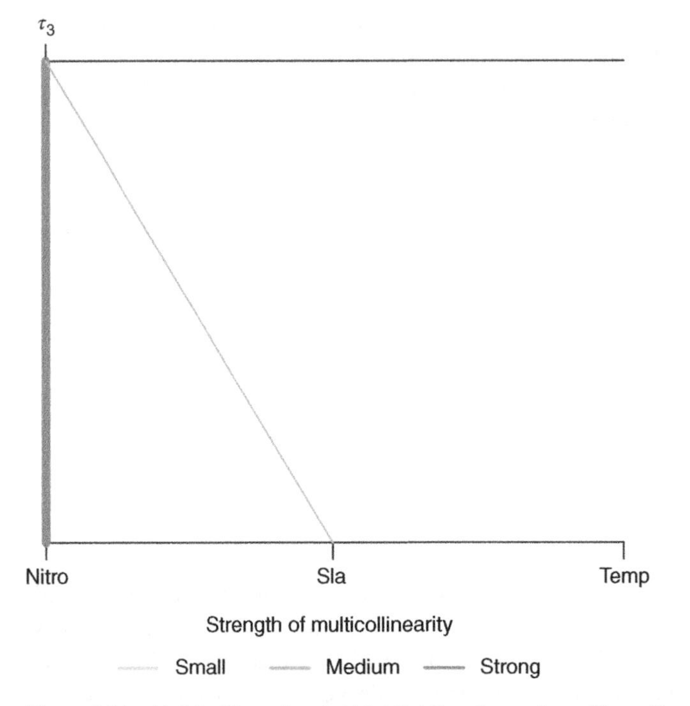

Figure 9.11 Multicollinearity plot highlighting the major collinearity-causing variables based on the `mcvis` method (R package *mcvis*). The τ statistic at the top horizontal axis quantifies the degree of collinearity in the dataset and its subscript gives the number of predictors (here three). The horizontal axis at the bottom accommodates all the predictors. The upper and lower axes are linked by the multicollinearity indices, represented as lines of different grey shades and widths. Darker and wider lines indicate larger multicollinearity indices, which allows easy identification of the predictors that contribute most to collinearity.

```
> vif(m5)
   temp      sla
2.03359 2.03359
```

However, if we are particularly interested in leaf nitrogen, we may also try to remove one of the other predictors with a large VIF value instead to see if that alleviates the problem.

```
## Refitted model without SLA
> m6 <- lm(leafresp ~ temp + nitro, data = lr3)
> summary(m6)

Call:
lm(formula = leafresp ~ temp + nitro, data = lr3)

Residuals:
     Min        1Q   Median        3Q       Max
-1.23424 -0.25152  0.07675  0.29891  0.68109
```

```
Coefficients:
            Estimate Std. Error t value Pr(>|t|)
(Intercept) -2.81892    0.43240  -6.519 3.18e-09 ***
temp         0.40697    0.03118  13.051  < 2e-16 ***
nitro        1.05600    0.40723   2.593    0.011 *
—
Signif. codes:  0 '***' 0.001 '**' 0.01 '*' 0.05 '.' 0.1 ' ' 1

Residual standard error: 0.4031 on 97 degrees of freedom
Multiple R-squared:  0.9348,      Adjusted R-squared:  0.9335
F-statistic: 695.3 on 2 and 97 DF,  p-value: < 2.2e-16


> vif(m6)
    temp    nitro
5.826937 5.826937
```

Retaining leaf nitrogen in the model at the expense of SLA yields plausible results but the VIFs > 5 suggest that multicollinearity is still an issue, albeit less serious than before.

There are more sophisticated methods to deal with multicollinearity issues than removing troublesome predictors, such as regularisation techniques where a penalty term is added to prevent overfitting and improve model performance, as used in lasso and ridge regression (see R package *glmnet*, Friedman et al. 2010). You can also perform a principal component analysis, which transforms the original explanatory variables into a set of uncorrelated variables (principal components) that can be used as predictors in a regression model. However, these approaches are beyond the scope of this introductory text.

9.2.2 Testing Interactions Among Predictors

So, our model `m5` with leaf temperature and SLA as predictors seems to be most suitable. But what if the response of leaf respiration to leaf temperature is influenced by SLA (or conversely the response to SLA could be modulated by leaf temperature). This is called an interaction and we can include an interaction term in our model to test whether it is statistically significant. Generally speaking, a significant interaction between two predictors implies that the relationship between one predictor and the response variable is influenced by the level of the other predictor. When two predictors are involved, we speak of a two-way interaction, but higher-order interactions are possible. Understanding higher-order interactions is important for accurately modelling complex relationships in the data. However, we should strive for a balance between model complexity and interpretability and be cautious about overfitting models, especially when dealing with limited sample sizes.

Interaction terms are implemented using the colon ':' symbol, linking two or more predictors. As this notation may become quite cumbersome with higher-order interactions, there is a shorthand notation using the asterisk '*' symbol, which includes the highest interaction and all nested interactions as well as the main terms (individual predictors). In the R code below, we first show the explicit formula notation followed by the shorthand alternative.

When specifying a model using `lm`, y ~ a + b + a:b is the same as y ~ a*b

```
## Linear model with a two-way interaction between leaf tempera-
ture and SLA

## Explicit formula notation
> m7 <- lm(leafresp ~ temp + sla + temp:sla, data = lr3)

## Same in shorthand notation
> m7 <- lm(leafresp ~ temp * sla, data = lr3)
> summary(m7)

Call:
lm(formula = leafresp ~ temp * sla, data = lr3)

Residuals:
     Min       1Q   Median       3Q      Max
-1.28864 -0.20788  0.04744  0.26326  0.80426

Coefficients:
             Estimate Std. Error t value Pr(>|t|)
(Intercept) -3.037547   0.572379  -5.307 7.17e-07 ***
temp         0.524153   0.041062  12.765  < 2e-16 ***
sla          0.121405   0.041642   2.915  0.00442 **
temp:sla    -0.005402   0.002564  -2.107  0.03773 *
—
Signif. codes:  0 '***' 0.001 '**' 0.01 '*' 0.05 '.' 0.1 ' ' 1

Residual standard error: 0.3962 on 96 degrees of freedom
Multiple R-squared:  0.9377,      Adjusted R-squared:  0.9357
F-statistic: 481.5 on 3 and 96 DF,  p-value: < 2.2e-16
```

The interaction between leaf temperature and SLA (also written as leaf temperature × SLA interaction) is statistically significant ($t = -2.107$, d$f = 96$, $P = 0.038$). This means that the effect of leaf temperature on leaf respiration is dependent on the level of SLA. And of course the reverse is also true: the effect of SLA on leaf respiration is influenced by the level of leaf temperature. If the interaction had not been statistically significant, we could have removed it and thus simplified the model.

In terms of our data analysis, this means that our model `m7`, which contains the main effects of leaf temperature and SLA and their interaction, is the most appropriate model for our data. The model diagnostic plots (not shown) do not indicate gross violations of the underlying assumptions, meaning that we can rely on this model when drawing statistical conclusions.

9.2.3 Model Selection and Comparison

Sometimes, several candidate models come into question for a dataset. How do we decide which model is best suited among a set of candidate models?

Perhaps the most robust and frequently used model selection tool is the ***Akaike Information Criterion*** (AIC; Burnham and Anderson 2004). The AIC seeks to strike a balance between model complexity and goodness-of-fit to determine the most appropriate model while avoiding overfitting. The AIC is calculated as follows:

$$\text{AIC} = -2 \times \log-\text{likelihood} + 2 \times \text{number of parameters}$$

AIC trades off explanatory power with complexity according to 'Principle of Parsimony', or Occam's razor (see Chapter 1).

The log-likelihood is a score that tells us how likely the model parameters and predictions are, given the observed data. The number of parameters refers to the number of parameters (coefficients) in the model.

The underlying formula indicates that the AIC penalises models with more parameters and thus favours simpler models, given equal explanatory power. The lower the AIC score, the more favourable this trade-off. This also applies in the negative range, i.e. the more negative the AIC score, the better the model performance. The difference in AIC scores (ΔAIC) is commonly used for model comparisons and the following rule of thumb is often seen in use:

- $\Delta\text{AIC} < 2$: Models perform equally well and are indistinguishable based on the AIC.
- $2 \geq \Delta\text{AIC} < 6$: Moderate evidence of a difference in model performance, lending support to the model with the lower AIC.
- $6 \geq \Delta\text{AIC} < 10$: Strong evidence that the model with the lower AIC performs better.
- $\Delta\text{AIC} \geq 10$: Very strong evidence that the model with the lower AIC is superior.

It is, however, important to note that the AIC is a relative measure that is only meaningful within a set of candidate models. The absolute AIC values have no interpretable meaning. The AIC is a model selection tool and not designed for hypothesis testing and, therefore, when reporting AIC scores or ΔAIC information, we must not speak about statistically significant differences.

We can create a set of candidate models and then use the `AIC` function to calculate the AIC for them or feed a list of these models into the more sophisticated `aictab` function (R package *AICcmodavg*, Mazerolle 2023) that creates a model selection table including the ΔAIC and other metrics. Here, we will use the neat `stepAIC` function which, starting from an initial (supplied) model, creates all possible simpler models and compares them with each other on an AIC basis to guide our model selection procedure (R package *MASS*, Venables and Ripley 2002). The term 'step' in the function name refers to the stepwise procedure of iteratively adding or removing predictors based on the AIC criterion.

```
## Load required package and rerun model m5
> library(MASS)

> m5 <- lm(leafresp ~ temp + sla, data = lr3)

> stepAIC(m5)
Start:  AIC=-178.75
leafresp ~ temp + sla

       Df Sum of Sq     RSS     AIC
<none>               15.763 -178.751
- sla   1     1.094  16.856 -174.044
- temp  1    95.470 111.233   14.646
```

```
Call:
lm(formula = leafresp ~ temp + sla, data = lr3)

Coefficients:
(Intercept)         temp          sla
   -1.90589      0.44650      0.03955
```

Only the upper half of the `stepAIC` output is relevant for model selection. The `<none>` indicates the initial model with an AIC value of 178.75. Removing SLA or leaf temperature from the model results in a less negative or even positive AIC score. In both instances, a $\Delta AIC > 2$ suggests that these predictors should be kept in the model.

We also have the option to specify the lower level of simplicity and the upper level of complexity of the candidate models fitted behind the scenes by the `stepAIC` function. In this way, we can test whether the inclusion of an interaction term is warranted. This is done by providing a list with right-hand side equations describing the predictor structure (a linear model equation without the response variable). We use our model `m5` with the main effects of leaf temperature and SLA as a starting point. Obviously, the most complex model would include the leaf temperature × SLA interaction. The simplest model only contains an intercept and no predictor variables. This model represents the mean response for all observations, assuming that the predictor variables have a value of zero. Such an intercept-only model is indicated by `~1` and represents the lower level default.

```
> stepAIC(m5, scope = list(upper = ~ temp * sla, lower = ~ 1))

Start:  AIC=-178.75
leafresp ~ temp + sla

             Df Sum of Sq     RSS      AIC
+ temp:sla    1     0.697  15.066 -181.272
<none>                     15.763 -178.751
- sla         1     1.094  16.856 -174.044
- temp        1    95.470 111.233   14.646

Step:  AIC=-181.27
leafresp ~ temp + sla + temp:sla

             Df Sum of Sq    RSS     AIC
<none>                    15.066 -181.27
- temp:sla    1   0.69666 15.763 -178.75
```

The addition of the leaf temperature × SLA interaction leads to a more negative AIC value (ΔAIC of 2.5), which lends support to this more complex model. We have seen earlier that the order in which the predictors are entered in the model formula matters. The function takes this into account by default, testing all possible predictor combinations.

`stepAIC` takes into account that the order in which parameters are supplied matters!

Despite the convenience of the `stepAIC` function, we must not forget to check the underlying assumptions of the most highly ranked model before we use it for statistical inference.

9.2.4 Variable Importance

In the presence of multiple significant predictors, the question naturally arises as to which of these are most relevant for model performance. The amount of explained variation given by the R^2 value appears to be a suitable assessment criterion but it includes the combined influence of all predictors. What we need though, is a decomposition of the model R^2 into its components, to give us the proportional contribution of each predictor to the total R^2, allowing an evaluation of their relative importance. Obviously, predictors with a larger partial R^2 would be considered more important.

Adding predictors sequentially to the regression model offers an intuitive approach since the associated increase in R^2 can be viewed as the contribution of the most recently added predictor. However, the sequence of the predictors in the model formula matters. To overcome this problem of the dependence on the order of the predictors, the `calc.relimp` function averages partial R^2 values over all possible permutations (a permutation is an arrangement of elements in a specific order).

For example, for a model with 4 predictors, there are 24 possible permutations ($4! = 4\times3\times2\times1 = 24$). For each permutation, the partial R^2 is calculated, and the function output provides the averages across the 24 permutations.

```
## Install and load package relaimpo to assess relative variable importance
> install.packages("relaimpo")
> library(relaimpo)

## Calculate relative importance as partial R-squares
> calc.relimp(m7, rela = F)

Response variable: leafresp
Total response variance: 2.442024
Analysis based on 100 observations

3 Regressors:
temp sla temp:sla
Proportion of variance explained by model: 93.77%
Metrics are not normalized (rela=FALSE).

Relative importance metrics:

                lmg
temp     0.662586103
sla      0.272213556
temp:sla 0.002881629
```

The most interesting information in the output is the relative importance metric called `'lmg'`, referring to **L**indeman, **M**erenda & **G**old's method (Lindeman et al. 1980). The `lmg` values represent the partial R^2 values (given as decimals) and sum up to the model R^2 of 93.77% ($0.662586103 + 0.272213556 + 0.002881629 = 0.9376813$). The output indicates that leaf temperature explains about two-thirds of the variation, followed by SLA explaining about 27% and quite a small contribution by the interaction.

9.2.5 Visualising Multiple Linear Regression Results

Ideally, we would plot all statistically significant linear relationships together. With two predictors, we can still visualise the relationship with the response variable in a 3D graph

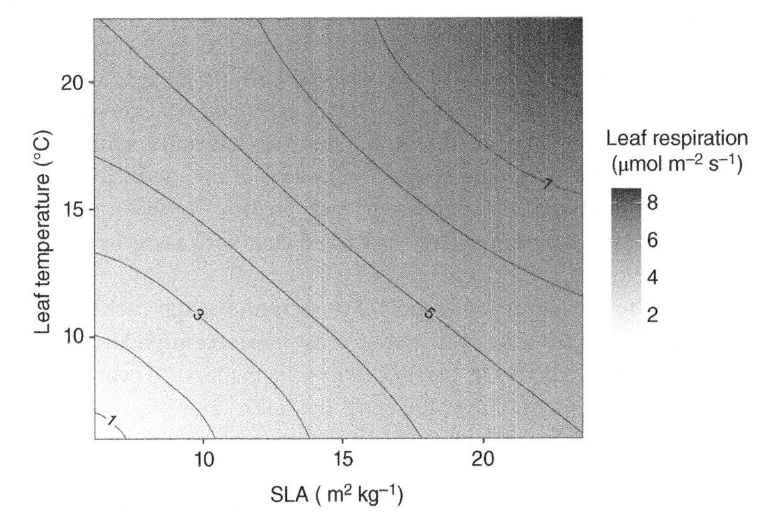

Figure 9.12 Grey scale contour plot of the predictions of the multiple regression model m7. The two predictors are plotted on the *y*- and *x*-axes, while the fitted values of the response variable are displayed as a colour gradient supplemented by contour lines connecting data points of the same value.

To visualise a regression with more than one predictor, use colour contour plots (two predictors) or added-variable plots.

including a regression plane instead of a regression line. However, most data visualisation experts discourage the use of 3D graphs for various valid reasons such as misleading perspectives, overplotting of points, blurring of patterns due to the extra dimension or poor accessibility for people with visual impairments, all of which complicate interpretation. A better alternative are colour contour plots (see Section 6.8.2), where the values of the response variable are displayed as a colour gradient, complemented by contour lines connecting points of the response variable with the same values (Figure 9.12, the R code for this figure is available on the book website, see also Chapter 6 for visualisation techniques).

The interpretation of the colour contour plot in Figure 9.12 is straightforward. The lowest leaf respiration rates around 1 μmol m^{-2} s^{-1} occur at leaf temperatures <12 °C and SLA values <10. High leaf respiration rates of 7 μmol m^{-2} s^{-1} or greater coincide with high leaf temperatures over 17 °C and SLA values over 18.

With more than two predictors we are looking at a multi-dimensional space, which becomes increasingly difficult for us to imagine, let alone plot. In those instances, a common way of visualising multiple linear regression results is using *added-variable plots* (also known as partial regression plots). These graphical tools are particularly useful in assessing the linear relationship between a certain predictor and the response variable while accounting for the effects of the remaining predictors (Figure 9.13). The term 'added-variable' relates to the idea that these plots can also be used to evaluate how the overall model performance changes by adding a new predictor variable. This helps to answer the question if the inclusion of a certain predictor improves the model, given the influence of the other predictor variables already present in the model. Of course, added-variable plots can already be used for multiple regression models with only two predictors as is the case here.

```
## Added-variable plots (partial regression plots)
> library(car)
> avPlot(m7, variable = "temp")
> avPlot(m7, variable = "sla")
```

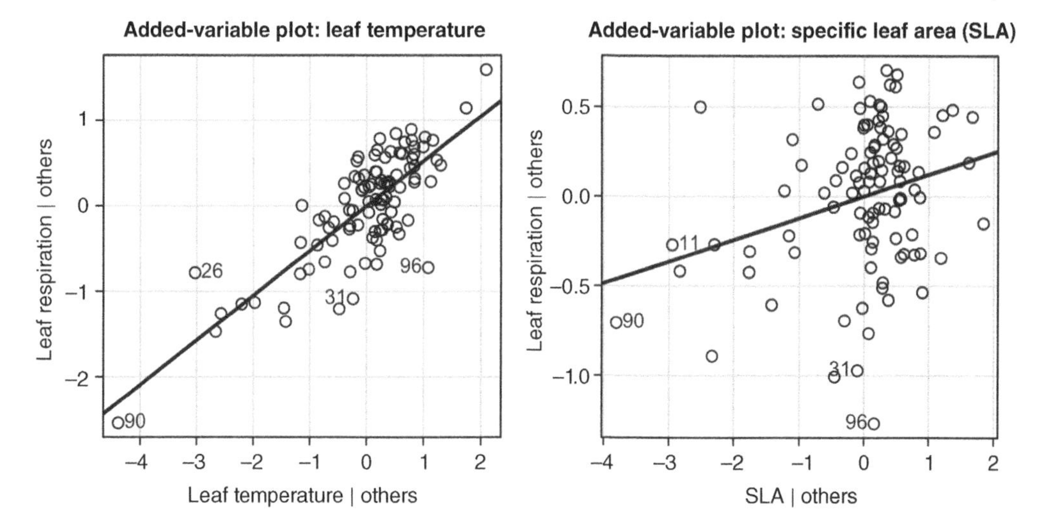

Figure 9.13 Added-variable plots showing how the selected predictor variable contributes to explaining the remaining variability in the response variable after accounting for the effects of the other predictors in the model. The '| others' bit reads as 'conditional on the other predictors' or 'given the other predictors'. By default, the `avPlot` function labels four points corresponding to the two observations with the largest residuals and the two observations with the greatest leverage.

The steeper the slope of a linear relationship in an added-variable plot, the greater the importance of the contribution of the selected predictor to the model given the effect of the remaining predictors. Very small, statistically non-significant slopes indicate that the selected predictor does not make a significant contribution to model performance and should thus not be included.

What stands out is that the values on both axes are centred around zero, which does not match the range of our input variables. So, where do these values come from and how do we interpret them? To understand this, we need to take a look under the hood. We will focus on leaf temperature as the predictor of interest (POI) (Figure 9.13, left panel). The trick of accounting for the effect of predictors other than the POI on the response variable is to run a regression model that includes all predictors except the POI. The variation in the response variable that is not explained by the other predictors is captured by the residuals, which are hence called *partial residuals*. Next, we run another regression model where we regress the POI (leaf temperature) against the remaining predictors (here SLA and the SLA × leaf temperature interaction) to assess how much of its variation is explained by the other predictors. The residuals from this regression represent the unique influence or contribution of the POI that is not explained by the other predictors. This approach allows us to examine the relationship between the POI and the response variable, while accounting for the effect of the remaining predictors.

In a nutshell, the residuals from regressing the response variable against all predictors other than the POI go on the y-axis, while the residuals from regressing the POI against all other predictors go on the x-axis. Well-behaved residuals scatter randomly around zero, which explains the axis ranges in Figure 9.13.

References

Burnham, K.P. and Anderson, D.R. (2004). Multimodel inference: understanding AIC and BIC in model selection. *Sociological Methods & Research* 33 (2): 261–304.

Fox, J. and Weisberg, S. (2019). *An R Companion to Applied Regression*, 3e. Thousand Oaks, CA: Sage https://socialsciences.mcmaster.ca/jfox/Books/Companion.

Friedman, J., Tibshirani, R., and Hastie, T. (2010). Regularization paths for generalized linear models via coordinate descent. *Journal of Statistical Software* 33 (1): 1–22. https://doi.org/10.18637/jss.v033.i01.

Lin, C., Wang, K., and Mueller, S. (2020). mcvis: a new framework for collinearity discovery, diagnostic and visualization. *Journal of Computational and Graphical Statistics* 30 (1): 125–132. https://doi.org/10.1080/10618600.2020.1779729.

Lindeman, R.H., Merenda, P.F., and Gold, R.Z. (1980). *Introduction to Bivariate and Multivariate Analysis*. Glenview, IL: Scott, Foresman and Company.

Lüdecke, D., Ben-Shachar, M.S., Patil, I. et al. (2021). performance: an R package for assessment, comparison and testing of statistical models. *Journal of Open Source Software* 6 (60): 3139. https://doi.org/10.21105/joss.03139.

Mazerolle, M.J. (2023). AICcmodavg: model selection and multimodel inference based on (Q)AIC(c). *R Package Version* 2 (3): 3. https://cran.r-project.org/package=AICcmodavg.

Pinheiro, J.C. and Bates, D.M. (2000). *Mixed-Effects Models in S and S-PLUS*, 528. New York: Springer https://doi.org/10.1007/b98882.

Pinheiro, J., Bates, D., and R Core Team (2023). nlme: linear and nonlinear mixed effects models. *R Package Version* 3: 1–162. https://CRAN.R-project.org/package=nlme.

R Core Team (2022). R: A Language and Environment for Statistical Computing In (Version 4.2.1) R Foundation for Statistical Computing. https://www.r-project.org/

Venables, W.N. and Ripley, B.D. (2002). *Modern Applied Statistics with S*, 4e. New York: Springer 498 p.

Zuur, A., Ieno, E.N., Walker, N. et al. (2009). *Mixed Effects Models and Extensions in Ecology with R*. New York: Springer.

10

One or More Categorical Predictors – Analysis of Variance

10.1 Comparing Groups

Analysis of variance (ANOVA) follows logically from the *t*-test (Chapter 5), where the difference in the mean between two groups is compared to the variation or the spread within the two groups. This type of analysis caters for cases where your categorical explanatory variable (sometimes called 'factor') has more than just two levels or groups. For example, while a *t*-test could be used to test the difference in sales between two packaging colours for a product, ANOVA can be used if you have more than just two colours to compare. Analogous to the *t*-test, we focus on the comparison between two metrics, one characterising the differences between the groups, and the other one within the groups. Intuitively, this makes sense, the mere differences between groups do not allow us to make a judgement, unless we know the contextual or background variation. If you look at the group means in Figure 10.1a, it is not possible to state whether group 1 differs from group 2 for example. We can see that the mean for group 2 is higher than the one for group 1, but this is meaningless without the knowledge of how the individual data points spread around the means. On the one hand, they could vary around the means as shown in Figure 10.1b, in which case the differences appear coincidental. We could also express this as '*the variance within the groups is high relative to the variance between the groups*'. On the other hand, if the measurements leading to the same mean values were to sit as shown in Figure 10.1c, we would be rather confident that true differences between the group means exist.

While a *t*-test compares two groups, ANOVA can compare several groups at the same time

10.2 Comparing Groups Numerically

Figure 10.1 provides a visual tool to answer the question 'Are the group means different?'. However, we need a way to quantify this numerically. Because we are concerned with the variation (once between mean values, and once between individual values within a group), we will use the variance as a metric (discussed in Chapter 2) to do so. A good way to compare the two is to look at the between-group to within-group variance ratio. If this ratio is large, this suggests that the groups are likely different, as the 'noise' (the variance within the groups) is small relative to the group differences (the signal, see Figure 10.1c). Note that this ratio is the equivalent of the signal-to-noise ratio discussed in Chapter 2.

The between-group variance is the signal, the within-group variance is the noise

R-ticulate: A Beginner's Guide to Data Analysis for Natural Scientists, First Edition.
Martin Bader and Sebastian Leuzinger.

Companion website: www.wiley.com/go/Bader

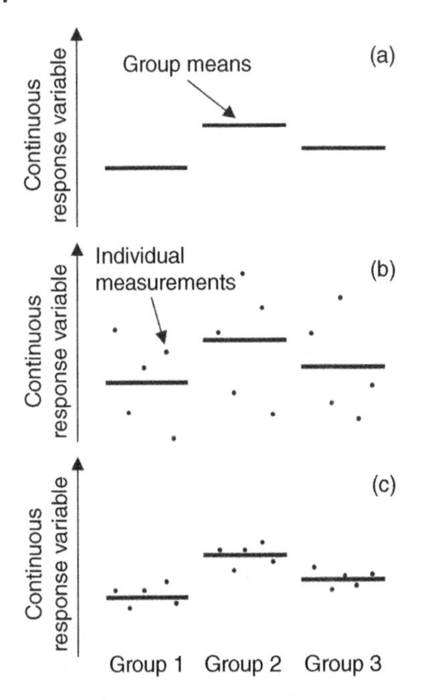

Figure 10.1 The fundamental idea of ANOVA. Differences between the same group means can originate from very different underlying data points. The numeric comparison between the variation among groups (horizontal lines) and the variation of measurements within one group (black dots) is the key for a probabilistic statement on whether the group differences are true. (a) Group means only, (b) same group means originating from data that spread widely around the means, and (c) again the same group means, this time from data points very close to their respective means.

Because the language to explain this inevitably becomes complicated, we will look at a numerical example. We first create a continuous response variable *y*, and a grouping variable *f*, the basic 'ingredients' for an ANOVA:

```
> y <- c(3, 4, 2, 1, 9, 7, 6, 8, 2, 3, 3, 1) # our response
  variable
> f <- gl(n = 3, k = 4, labels = c("A", "B", "C")) #
  generate a factor with 3 levels and 4 replicates per level
  using gl()
```

The dataset is visualised in Figure 10.2a in the form of a boxplot, giving us a rough idea of how the groups A, B, and C differ in their *y* values. The ultimate question is whether the *y* measurements differ significantly by group. You can see from Figure 10.2c that in order to compute the within-group variance, we need to subtract the *y* values from their group mean (vertical lines in Figure 10.2c). For the between-group variance, we need to subtract the group means from the grand mean (overall mean of *y*) for every data point (represented by the vertical lines in Figure 10.2d):

If you have k groups and n replicates per group, you have $k - 1$ degrees of freedom for the between-group variance (numerator), and $n - k$ degrees of freedom for the within group (or error) variance (denominator).

```
> means <- tapply(y, f, mean) # calculate group means
> groupmeans <- rep (means, each = 4) # extend to vector
  of length 12
> grandmean <- mean(y)
```

Note that we need to repeat each group mean four times in order to provide a group mean for each observation. With the raw observations (*y*), the grand mean, and the group means, we are now equipped to calculate the between- and within-group variances. Why do we divide the between-group sum of squares by 2, but the within-group sum of squares by 9? Two and 9 are the respective '*degrees of freedom*'. Loosely spoken, this is because for the between-group variance, 2 group means are 'free to vary' (the third one is locked in by the overall mean), and, analogously, for the within-group variance, we can choose three values per group freely (so 9 in total, the fourth one is locked in by the calculations of the respective group mean). See Chapter 2 for an explanation of degrees of freedom. We therefore say that the variance ratio

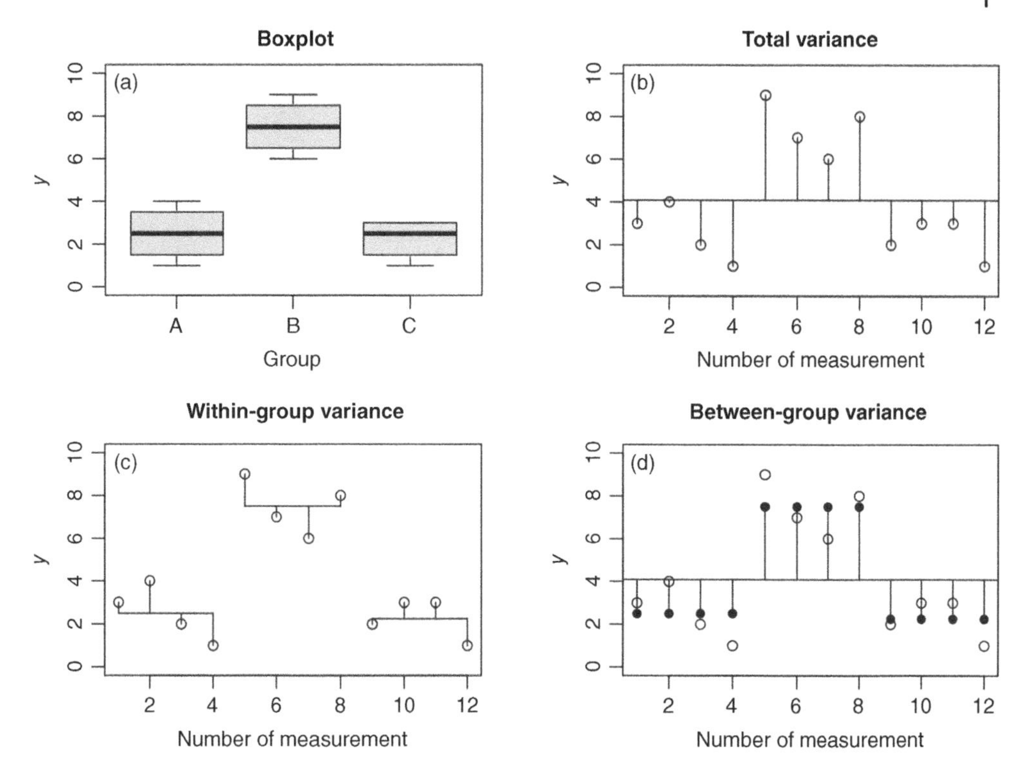

Figure 10.2 A simple example of an ANOVA visualised. (a) Boxplot showing the differences of *y* between the groups. (b) The total variance in the dataset, the horizontal line is the grand mean. (c) The within-group variance (the differences between the individual data points and their respective group mean). (d) The between-group variance (the differences between the group means and the grand mean, shown for every observation separately).

in our example follows an $F_{2,9}$ distribution (say 'an *F* distribution with 2 degrees of freedom in the numerator, and 9 degrees of freedom in the denominator').

```
> between <- sum((mean(y) - groupmeans)^2)/2
> between
[1] 35.08333
> within <- sum((y - groupmeans)^2)/9
> within
[1] 1.416667
```

We can already see that the between-group variance is much larger than the within-group variance, simply by looking at the average length of the vertical lines in Figure 10.2c vs. d. This is confirmed by their exact calculation (see above code), and suggests that the groups indeed differ in *y*, lending support to the qualitative observation we make from the boxplot in Figure 10.2a.

In order to make a quantitative conclusion whether our response variable differs significantly between the three groups, we now divide the between-group by the within-group variance, in our case $35.08/1.42 = 24.70$. The larger the value, the higher the signal and the lower the noise, and therefore the more likely there is a true difference in *y* between groups. But at what point can we call this difference significant in a statistical sense? Much like with the earlier tests we conducted, the argument goes: 'If we ran multiple ANOVAs using random numbers, how often would the variance ratio be *as high (or higher)* as in our case (i.e. 24.7 or higher)? Well, let us simulate this:

```
> ratio <- NULL # an object to 'glue' the ratios onto
> for (i in 1:1000) { # for the use of loops, see Box 10.1
+   y1 <- rnorm(n = 12) # create random normally distributed
   numbers
+   means <- tapply(y1, f, mean) # calculate group means
+   groupmeans <- rep(means, each = 4)
+   between <- sum((mean(y1) - groupmeans)^2)/2
+   within <- sum((y1 - groupmeans)^2)/9
+   ratio <- append(ratio, between/within) # appends the
   calculated ratios
+ }
> hist(ratio) # see Figure 10.3
```

Box 10.1 Programming Loops

Loops are a handy tool if you need to execute functions or algorithms many, sometimes thousands of times. They are often used to simulate something, but also in figures, for example to draw multiple error bars at the same time. Often, they can be avoided using more efficient (*but less intuitive!*) vectorised programming tools, but with today's computing power, this argument has somewhat lost traction. The most widely used loop command is `for`. Simpler versions are `repeat` or `while`, which repeat the specified algorithm infinitely. The function `break` causes the loop to end instantly, and it is often used to exit the loop. The following example illustrates this:

```
> i <- 0 # initialisation
> repeat {
+   i <- i + 1        # add 1 to i
+   if(i == 3) break # stop if i = 3
+ }
> i
[1] 3
```

The above code also introduces the use of `if`, which we will not explain further here. The more common loop is the classic `for` loop. An example that you can play with explains it best:

```
x <- c(3, 6, 4, 8)
for(i in x) { # i takes on the values 3, 6, 4, and then 8
  print(i^2)
}
[1] 9
[1] 36
[1] 16
[1] 64
```

It is sometimes easier to understand a loop if you put yourself in R's position. In the above example, you simply run the algorithm (which here says print the square root of i). What values does i take on? You can see that in brackets above: i runs through all values in *x* (hence i *in* x), so 3, 6, 4, and 8. Of course this job could have been done much more simply:

```
> c(3, 6, 4, 8)^2
[1]  9 36 16 64
```

Imagine however that you have 100 data files that are automatically created: if you wanted to import them into R, you would have to write 100 lines, one for every file (or undergo a major copy/paste job). To simulate this, we first write 100 files into your work directory, then we read them again and paste them all together so you end up with one single object in R including all the data. The files each contain random normally distributed numbers.

```
> for (i in 1:100){
+ write.table(rnorm(100), file = paste("testfile", i, ".txt",
   sep = "")) # the sep argument is to avoid an empty space
+}
```

Check in your work directory or with `list.files` (no arguments) to see whether the files have been written. Imagine now that those are the 100 files you want to import into R and paste together to create one single object including all data.

```
> master <- NULL # initialise the master
> for(i in 1:100) # add all 100 files, attaching them with
   rbind
+ {
+ rbind(master, read.table(file = paste("testfile", i, ".txt",
+ sep = ""))) -> master
+ }
> dim(master)
[1] 10000 1
```

You will immediately agree that this could save a lot of work. If you want to read the files and assign them to R objects individually, use `assign`:

```
for (i in 1:100) {
+ assign(paste("table", i, sep = ""), read.table(file =
+ paste("testfile", i, ".txt",
+ sep = "")))
+ }
```

If you need to evaluate text that you paste together in a loop, you will need `eval` combined with `as.name`. This is relatively simple but unfortunately rarely mentioned:

```
> for (i in 1:100) {
+ plot(eval(as.name(paste("table", i, sep = ""))))
+ }
```

You can see from Figure 10.3 that the signal-to-noise ratio is high, in fact *unusually* high when compared to the average ratios from our 1000 analyses using random numbers.

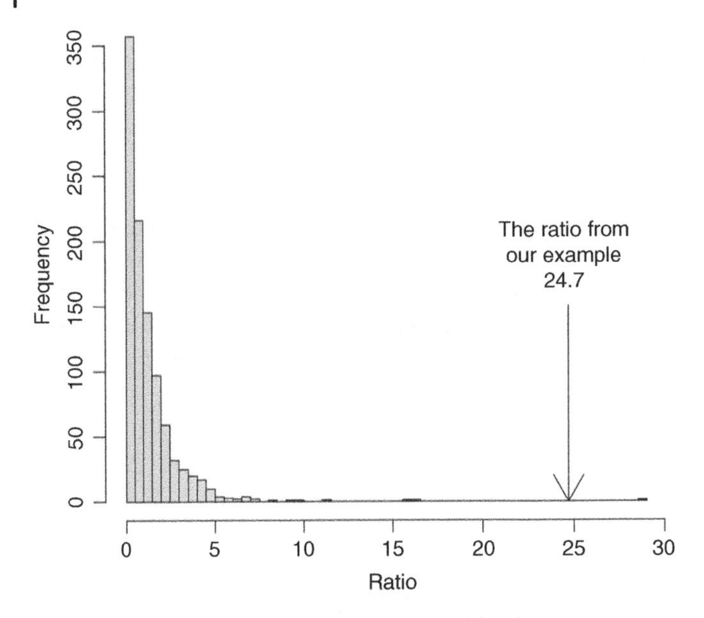

Figure 10.3 Histogram of random between- to within-group ratios of 1000 'handmade' ANOVAs. You can see that the most common ratios based on random numbers sit between zero and five, 'our' ratio is much higher and therefore suggests that the differences between the groups are real.

We now need a more formal way of testing our variance ratios. We therefore ask what distribution random variance ratios follow. From Figure 10.3, you can see that the distribution is certainly not normal, for one because the distribution is clearly not symmetrical. Variance ratios follow an F-distribution. Just like with the t-distribution, there is a dependency on the degrees of freedom, namely those of the between-group and that of the within-group variance as we have seen above.

Now we did not have to do this simulation by hand just to find out how rare our F-value, i.e. our variance ratio is! Of course R provides you with the whole suite of probability functions for the F-distribution, just like for the other distributions we have already encountered (see Chapter 3):

```
> rf(n = 100, df1 = 2, df2 = 9) # producing random F values
> qf(p = .5, df1 = 2, df2 = 9) # calculating a quantile for an F
   distribution
[1] 0.7493807
> pf(q = 24.76, df1 = 2, df2 = 9) # calculating a probability
[1] 0.9997806
```

The first line produces random $F_{2,9}$-values, the second one gives the median $F_{2,9}$-value, and the last line shows the probability to hit an $F_{2,9}$-value that is lower than 24.7. However, because we are interested in the probability of finding an equally large or larger test statistic only by chance, we need to specify the upper tail (right-hand side) of the distribution in the `pf` function by setting the argument `lower.tail = F` (see Section 3.4.2). The probability to score higher than our F-value of 24.7 is the P-value of our 'handmade' ANOVA!

```
## Calculating the probability in Figure 10.3
> pf(q = 24.76, df1 = 2, df2 = 9, lower.tail = F)
[1] 0.0002193925
```

10.3 One-way ANOVA Using R

After we have had a good look at how this ANOVA engine works, we now deserve to sit back, relax, press the button, and let R do the work. In fact, the last few pages can be condensed to a couple of simple R commands:

```
## ANOVA of y according to grouping in f (variables defined
   earlier)
> m1 <- aov(y ~ f)
> summary(m1) # summary output
            Df Sum Sq Mean Sq F value    Pr(>F)
f            2  70.17   35.08   24.77 0.000219 ***
Residuals    9  12.75    1.42
—
Signif. codes:  0 '***' 0.001 '**' 0.01 '*' 0.05 '.' 0.1 ' ' 1
```

Because this is a frequentist statistical test, we need a null hypothesis (see Chapter 5). Extending the null hypothesis of a two-sample test, in the case where you have several groups, it becomes: 'There is no difference in the response variable between the groups.'

The P-value (designated 'Pr(>F)' in the output) is therefore interpreted as the probability of finding an equally or more extreme test statistic under the assumption that the null hypothesis is true. Thus, if this probability is low, usually lower than the agreed upon 5% threshold ($\alpha = 0.05$), we reject the null hypothesis and accept the alternative hypothesis, which states that the grouping variable has a significant effect on our response variable. The detailed R output is dissected in Figure 10.4.

Note that the very same result can be achieved by using the function `lm` as an ANOVA is nothing more than a linear model that is fed categorical predictors. To obtain the classic ANOVA table as shown in Figure 10.4, you use the function `anova`:

```
> m1 <- lm(y ~ f)
> anova(m1)
```

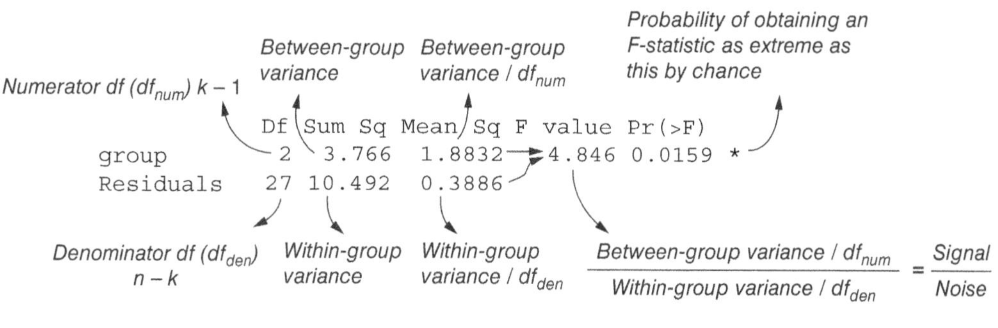

Figure 10.4 A detailed explanation of the standard ANOVA output in R.

This produces exactly the same result, and it is important not to confound the two functions `aov` and `anova`, the latter is to be interpreted as 'show me the ANOVA table of the linear model'.

We now have a very thorough understanding of the mechanics of ANOVA, but some important questions remain, e.g. what are the model assumptions and how can we verify them? What if they are violated? Between which groups exactly are potential differences actually significant? What if we have more than one categorical predictor? The remainder of this chapter will look at these questions.

10.4 Checking for the Model Assumptions

The validity of our ANOVA output rests, just like with the previous types of frequentist analyses, on verification of the underlying model assumptions. For ANOVA, these are:

- Normality of the residuals
- Homogeneity of variance (or homoscedasticity, i.e. approximately equal variation of residuals within the different groups)
- Independence of samples

While you can extract the residuals from the model output quite easily using `resid(m1)` it is easiest to look at the four diagnostic plots using `plot(m1)`, `m1` being the model output as defined earlier:

```
> par(mfrow = c(2, 2)) # because there are four plots to come
> plot(m1) # see Figure 10.5
```

The four standard diagnostic plots (Figure 10.5) allow us to check for the first two assumptions and, analogous to the diagnostic plots of a linear regression analysis seen earlier, a means of pinpointing influential data points (outliers). The third assumption (independence of observations) has to be considered at the stage of planning the experiment. Proper randomisation and replication (treated in Chapter 4) will help, and as demonstrated there, we can see how unintended and unnoticed dependence of subjects of the same group will inflate type I and II errors in ANOVA. What if one or more of the assumptions are violated? Apart from transformations (Chapter 8), we offer a glimpse at more advanced models that can relax both the homogeneity of variance as well as the 'normality of residuals' assumption (see Chapter 12). Another robust (distribution-free) method is the Kruskal–Wallis test briefly introduced in Section 10.7.

10.5 *Post Hoc* Comparisons

In a *t*-test, it is obvious that a significant *P*-value means that a real difference between the tested groups is likely. But what exactly does a significant *P*-value mean in an ANOVA that tests several groups? Which groups differ from one another? In fact, a significant *P*-value here is not very informative at all, it simply means that *at least one group differs significantly from at least one other group*. Of course, we want to know *which* groups differ. For this, we need a 'post hoc' test, *post hoc* simply meaning that we use this test *after* the actual test, the ANOVA.

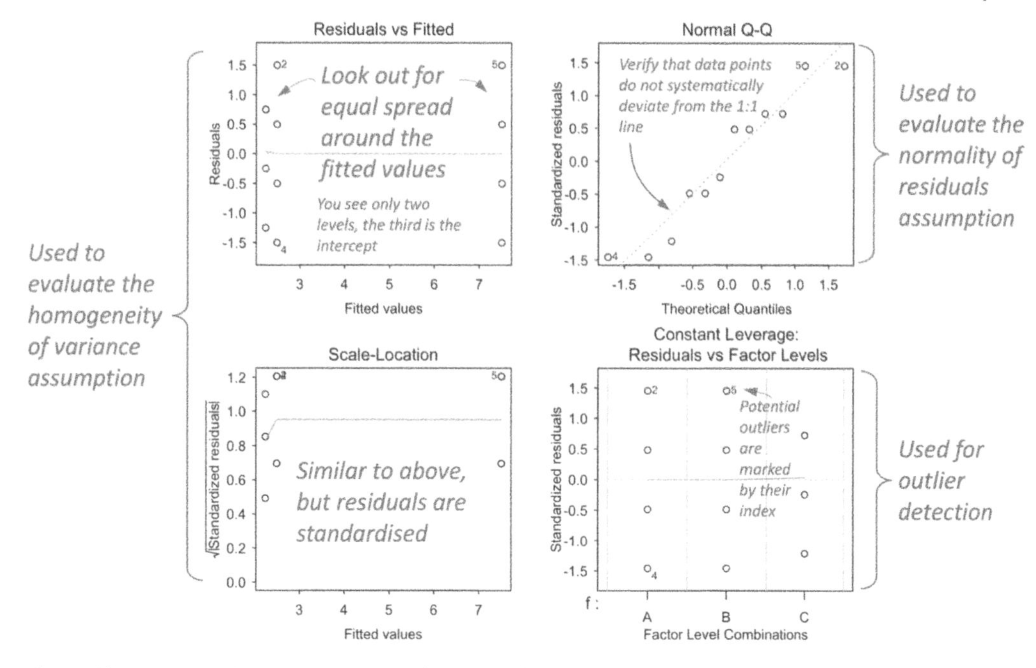

Figure 10.5 Model diagnostic plots showing the residuals vs. fitted (predicted) values (top left) – for assessing the variance homogeneity (homoscedasticity) criterion. An approximately equal spread around the fitted values suggests no heterogeneity of variance. The quantile-quantile plot (top right) contrasts the quantiles of the residuals with the corresponding quantiles of a standard normal distribution. If the sample distribution approximates a normal distribution, then the points tend to fall on the 1:1 line. The scale-location plot (bottom left) is similar to the first plot but uses the square root of the standardised residuals instead. The bottom right plot identifies potential outliers (labelled with their index).

The most basic and classic ANOVA *post hoc* test is 'Tukey's honestly significant difference' or Tukey's HSD test. Assuming we have three groups, like in our simple example from earlier, we would like to see three *P*-values to test whether 'A' differs from 'B', whether 'A' differs from 'C', and whether 'B' differs from 'C'. The number of comparisons quickly increases with the number of groups of course. So why can we not just use *t*-tests to compare the groups? Apart from the fact that it might become laborious once we have many groups, we run into a problem termed 'multiple comparisons'. With only three groups, the problem might be minute, but imagine what happens if you have an increasing number of groups to compare. Simply by chance, the probability of two groups differing will increase! Consider for example comparing age classes – if you increase the number of age classes you compare (e.g. by decreasing the size of the age classes), you necessarily increase the chance of finding that at least one group differs from any other one. We will demonstrate this quickly, and then return to Tukey's HSD test.

If you conduct multiple comparisons using the same dataset, the probability of finding significantly different pairs by pure chance increases with every comparison you make!

We will create 1000 numbers that follow a normal distribution, and then run random *t*-tests with a sample size of 10. First we run just 10, then 50, and then 100 tests. We will save the *P*-values for these tests in objects called `few`, `more`, and `many`. Then we test how many significant *P*-values we obtain in each case using the functions `length` and `which`. You will observe that we will systematically see more significant *P*-values if we use more tests! This is the problem with multiple comparisons. The exercise gives us a nice opportunity to apply a loop as described in Box 10.1. Note that the index `i` is not used in the algorithm itself, but it determines how many times we run through the loop.

```
> ## Create a population to randomly sample from
> pop <- rnorm(1000)
>
> few <- NULL # initiate a 'few', 'more', and 'many' object
> more <- NULL
> many <- NULL
>
> ## Conduct 10 random t-tests with a sample size of 10
> for (i in 1:10) {
+   few <- append(few, t.test(sample(pop, 10), sample
  (pop, 10))$p.value)
+ }
> length(which(few < 0.05)) # how many are significant?
[1] 0
>
> ## Conduct 50 random t-tests with a sample size of 10
> for (i in 1:50) {
+   more <- append(more, t.test(sample(pop, 10), sample
  (pop, 10))$p.value)
+ }
> length(which(more < 0.05)) # how many are significant?
[1] 4
>
> ## Conduct 100 random t-tests with a sample size of 10
> for (i in 1:100) {
+   many <- append(many, t.test(sample(pop, 10), sample
  (pop, 10))$p.value)
+ }
> length(which(many < 0.05)) # how many are significant?
[1] 8
```

With 10 tests, we find no groups that differ significantly, with 50 we find 4, and with 100 comparisons, we find 8 *t*-tests that turn out significant, simply by chance! Ok, this teaches us that we cannot just blindly compare without considering this purely probabilistic effect of obtaining an increasing number of significant *P*-values. Tukey's HSD test accounts for exactly that, and it is easy to use for simple one-way ANOVAs where there are no interactions. Let us again use `f` and `y` as defined earlier in the chapter.

```
> m1 <- aov(y ~ f)
> TukeyHSD(m1)
  Tukey multiple comparisons of means
    95% family-wise confidence level

Fit: aov(formula = y ~ f)

$f
```

```
      diff        lwr       upr     p adj
B-A  5.00  2.650177  7.349823 0.0005705
C-A -0.25 -2.599823  2.099823 0.9527766
C-B -5.25 -7.599823 -2.900177 0.0003989
```

We can easily see that while 'A' is different from 'B', and 'C' is different from 'B', the difference between 'A' and 'C', is not significant, this is visually easily confirmed in Figure 10.2a. The function further returns a lower and upper limit for the difference, as well as the absolute difference between means.

10.6 Two-way ANOVA and Interactions

A two-way ANOVA is an ANOVA with two categorical predictors. The novel element with two or more predictors is the 'interaction term' between the predictors. This adds complexity to the interpretation of an ANOVA. We have already briefly touched on interactions in Chapters 6 and 9, and we will come back to the concept in Chapter 11. To get started, we use a simple dataset we created in Chapter 6:

```
> set.seed(9) # a made up dataset
> d1 <- data.frame(yield = rnorm(40, mean = 20),
           fert = factor(rep(c('N', 'ctrl'), times = 20)),
           water = factor(rep(c('water', 'ctrl'), each = 20))
           )
```

A farmer wants to test the effect of fertiliser (nitrogen, 'N') and irrigation ('water') on their crop. The variable 'yield' is the response, and 'fert' and 'water' are the two predictors. If the two predictors are 'orthogonal', i.e. each of the predictor combinations is replicated, then we can ask three questions: (i) what is the effect of the fertiliser treatment on crop yield? (ii) What is the effect of irrigation on crop yield? And, (iii), what is the effect of the interaction of fertiliser and irrigation on crop yield?

An 'orthogonal' design is when each factor combination occurs more than once, i.e. is replicated

It is important to truly understand the concept of an orthogonal design. The simple question to ascertain is: 'do we have at least three measurements (here yield) at every possible factor combination? (Here 'water-N', 'water-control', 'N-control', and 'control-control')?' In our case, we indeed do, in fact we have 10 replications at each factor combination:

```
> table(d1[, c('fert', 'water')])
      water
fert    ctrl water
  ctrl    10    10
  N       10    10
```

So far so good, now let us return to the posed questions: To answer them we conduct a two-way ANOVA, which is no more difficult than conducting a one-way ANOVA. The asterisk '*' between the two predictors indicates that we want to include the interaction term. You could just test for the main effects using `aov(yield ~ fert + water, data = d1)`, but it would not make sense to exclude the interaction term at the start. Using the asterisk symbol as a short-hand notation for the two main effects and their interaction is equivalent to explicitly including the interaction term: `fert + water + fert:water`

```
> m2 <- aov(yield ~ fert*water, data = d1)
> summary(m2)
             Df Sum Sq Mean Sq F value Pr(>F)
fert          1  4.107   4.107   5.505 0.0246 *
water         1  1.225   1.225   1.643 0.2082
fert:water    1  3.169   3.169   4.247 0.0466 *
Residuals    36 26.858   0.746
- - -
Signif. codes:  0 '***' 0.001 '**' 0.01 '*' 0.05 '.' 0.1 ' ' 1
```

Are the two brothers good together? Yes, but only when they are with grandma! This is the equivalent of a significant interaction term – the question whether they are good together cannot be answered properly without specifying in whose care they are!

The interpretation of the output is analogous to what is shown in Figure 10.4, but what is new is the interaction term. You can see that the *P*-value of the interaction term is significant. What does that mean? It is perhaps most intuitive to look at the corresponding interaction plot (see Figure 6.7 in Chapter 6). You can see that there is an effect of fertiliser on crop yield *but only in the watered treatment*. This is exactly what a significant interaction term means – the effect of one predictor on the response depends on another predictor. So how do we respond to the farmer's question 'does fertiliser/water addition help to increase yield?'. The answer must be: 'fertiliser only helps in combination with irrigation'. What about the significant main effect of fertiliser? It has no significance any more, as stating that 'fertiliser promotes crop yield' is both wrong and true – it depends on another predictor, namely water addition. As a general rule – the moment an interaction term is significant, you can ignore the *P*-values of the underlying main effects – their interpretation is simply not possible.

In Chapter 11, we will get to know the package `emmeans`, which will make *post hoc* comparisons even easier, particularly if there are higher order interactions, and a mix between continuous and categorical predictors.

10.7 What If the Model Assumptions Are Violated?

Again, you will find that more often than not, model assumptions are violated. The options in the case of ANOVA are

- Transformations as described in Chapter 8. These are now far less common, and not well suited for ANOVA with multiple predictors, as the interpretation of significant interaction terms becomes difficult
- Add explanatory (co-)variables to extract variation, see Chapter 11 on ANCOVA
- If the assumptions of normality and/or homogeneity are violated only mildly, still perform the test, but interpret the *P*-values more conservatively
- Use a more elaborate model that can relax the assumptions (see Chapter 12)

A non-parametric ANOVA cannot account for interaction terms

A further alternative is the Kruskal–Wallis test. Just like the Wilcoxon rank sum test (see Section 5.2.2), it is based on ranks, and therefore free from any assumptions relating to the distribution of the residuals. We have to bear in mind that it is a much more simplistic test, because you sacrifice a lot of information (namely the absolute values of your response variable). The Kruskal–Wallis test is also unable to deal with interactions, for the very same reasons that data transformations affect the interpretation of interactions: they are affected by the absolute values of the response. Nevertheless, a Kruskal–Wallis-type ANOVA can be useful for small sample sizes and simple designs without interactions. To illustrate its use, we will analyse a fabricated dataset using both a standard ANOVA and the non-parametric alternative.

```
> ## Set a random seed so you can reproduce the example
> set.seed(5)
> ## Use non-normal values for the response (F-distribution)
> resp <- rf(32, 1, 2)
> ## Use a categorical predictor with four levels and 8
   replications
> pred <- gl(4, 8, labels = letters[1:4])
> d2 <-  data.frame(resp, pred) # create a data frame
> ## This plot using a log scale is not shown here:
> boxplot(log(d2$resp) ~ d2$pred)
```

We first use a standard ANOVA, which does not pick up a significant difference between the four groups:

```
> summary(aov(resp ~ pred, data = d2))
            Df Sum Sq Mean Sq F value Pr(>F)
pred         3   8087    2696   1.436  0.253
Residuals   28  52579    1878
```

The Kruskal–Wallis test on the other hand, based on ranks rather than absolute values, produces a significant P-value:

```
> kruskal.test(resp ~ pred, data = d2)

       Kruskal-Wallis rank sum test

data:  resp by pred
Kruskal-Wallis chi-squared = 8.4886, df = 3, p-value = 0.03692
```

If we want to follow up with a *post hoc* test to find out which groups are different, we use the Dunn test after installing and loading the package FSA (Dinno 2017):

```
> library(FSA)
> dunnTest(resp ~ pred, data = d2)
Dunn (1964) Kruskal-Wallis multiple comparison
  p-values adjusted with the Holm method.

  Comparison          Z    P.unadj      P.adj
1      a - b  0.6929023 0.48837085 0.97674170
2      a - c -1.3858047 0.16580656 0.49741968
3      b - c -2.0787070 0.03764429 0.18822145
4      a - d -1.8655063 0.06211047 0.24844190
5      b - d -2.5584086 0.01051525 0.06309148
6      c - d -0.4797016 0.63143958 0.63143958
```

This shows that only between groups ‘b’ and ‘d’, there is a (marginally) significant difference. Note how the P-values change from the ‘unadjusted’ to the ‘adjusted’ column, correcting for the increased probability of finding significant differences by chance (see Section 10.5).

Reference

Dinno, A. (2017). dunn.test: Dunn's Test of Multiple Comparisons Using Rank Sums. R package version 1.3.5, https://CRAN.R-project.org/package=dunn.test.

11

Analysis of Covariance (ANCOVA)

When we have a continuous response variable and a mix of continuous and categorical explanatory variables (at least one of each), we use a so-called *analysis of covariance* model (**ANCOVA**). In R, this type of model can be implemented using the familiar `lm` function. Let us explore this type of analysis with a dataset consisting of the growth rate (`growth`, μm day^{-1}) of three different pathogenic fungal species as a function of fungicide concentration (`conc`, mg l^{-1}). The data are stored in the provided dataset called `pathfung`. As a 'sanity check', we first look at the structure (`str`) and the summary (`summary`) of the data.

```
## Read in data
> pathfung <- read.csv("pathogenic_fungi.csv")

## Check structure and summary
> str(pathfung)

'data.frame':       54 obs. of  3 variables:
 $ conc   : int  10 8 6 4 2 0 10 8 6 4 ...
 $ growth : num  37.8 85.3 111.4 147.3 236.4 ...
 $ species: chr  "A" "A" "A" "A" ...
```

The `str` output shows fungicide concentration as an integer and fungal growth as a numeric (continuous) variable, which is fine. However, species is currently coded as character but since it is a categorical variable (factor), it should be coded as a factor (see Box 1.4 in Chapter 1). We will do this right away.

```
## Recode species as factor
> pathfung$species <- as.factor(pathfung$species)
```

It is a good idea to rerun the `str(pathfung)` command at this stage to ensure that `species` is now a factor.

The summary of this dataset flags no issues. The concentration ranges from 0 to 10 mg l^{-1}, the growth rate seems to make sense (no negative or exceedingly high values) and we can see that there are 18 observations per fungal species, so no further data wrangling action required.

R-ticulate: A Beginner's Guide to Data Analysis for Natural Scientists, First Edition.
Martin Bader and Sebastian Leuzinger.

Companion website: www.wiley.com/go/Bader

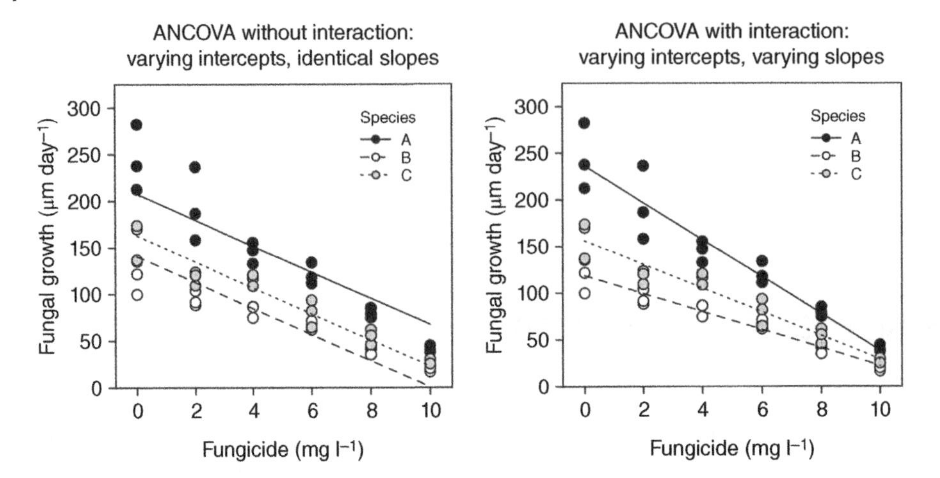

Figure 11.1 ANCOVA fit without interaction (left) and fit of an ANCOVA with interaction allowing separate slopes for each fungal species (right).

```
> summary(pathfung)

      conc          growth           species
 Min.   : 0    Min.   : 17.05     A:18
 1st Qu.: 2    1st Qu.: 57.28     B:18
 Median : 5    Median : 90.04     C:18
 Mean   : 5    Mean   :100.32
 3rd Qu.: 8    3rd Qu.:130.63
 Max.   :10    Max.   :282.29
```

Next, we plot the data and fit an ANCOVA model without interaction term, get the model predictions, and add those as regression lines to the plot (Figure 11.1).

```
> plot(growth ~ conc, data = pathfung, pch = 21,
     bg = c("black", "white", "grey")[pathfung$spec]) # species colour coding
```

Similar to multiple regression models, an ANCOVA without interaction requires that the predictor (explanatory) variables show up on the right-hand side of the squiggly line (tilde symbol) separated by a plus sign. We start with a simple ANCOVA model that only considers the main effects of fungicide concentration and fungal species but not their interaction. This model assumes that the three fungal species share the same slope but the intercepts are allowed to vary ('the regression lines may hover at different heights').

```
## ANCOVA without interaction
> m1 <- lm(growth ~ conc + species, data = pathfung)

> summary(m1)
Call:
lm(formula = growth ~ conc + species, data = pathfung)

Residuals:
    Min      1Q  Median      3Q     Max
-40.822 -14.317   2.426  10.705  74.753
```

```
                                                        Std.      t
                          Coefficients Estimate    Error    value  Pr(>|t|)
    Intercept (Int.) speciesA ⬅ (Intercept) 207.5348 6.4663  32.095  < 2e-16 ***
                 Common slope ⬅ conc        -14.0050 0.8348 -16.777  < 2e-16 ***
Int. speciesB - int. speciesA ⬅ speciesB    -66.8433 6.9844  -9.570 6.90e-13 ***
Int. speciesC - int. speciesA ⬅ speciesC    -44.7144 6.9844  -6.402 5.17e-08 ***

---
Signif. codes:  0 '***' 0.001 '**' 0.01 '*' 0.05 '.' 0.1 ' ' 1

Residual standard error: 20.95 on 50 degrees of freedom
Multiple R-squared:  0.8828,     Adjusted R-squared:  0.8757
F-statistic: 125.5 on 3 and 50 DF,  p-value: < 2.2e-16


## Get model predictions (model fit) and add them to the data frame
> pathfung$fits <- predict(m1)

## Add the model predictions (fitted lines) to the plot
> lines(fits ~ conc, data = pathfung[pathfung$species == "A", ], lty = 1)
> lines(fits ~ conc, data = pathfung[pathfung$species == "B", ], lty = 2)
> lines(fits ~ conc, data = pathfung[pathfung$species == "C", ], lty = 3)

## Advanced coding using a loop for adding the fitted lines
> for (i in 1:3) {
+   lines(fits ~ conc, data = pathfung[pathfung$species ==
          levels(pathfung$species)[i], ], lty = c(1:3)[i])
+ }
```

11.1 Interpreting ANCOVA Results

The interpretation of an ANCOVA output is not straightforward and thus often leads to confusion. To facilitate understanding, we annotate the model summary output and explain the meaning of the coefficients in detail in the following text.

The `intercept` is the estimate of the response (y-value) when the predictor (x-value) is zero. In the model summary output, the intercept term represents the intercept of species A, i.e. the growth rate of species A in the absence of the fungicide (`conc` = 0), which is about 208 mm day^{-1}. The `conc` coefficient gives the *common slope* for all species, which is roughly −14, indicating that fungal growth decreases by 14 mm day^{-1} with a one unit increase in fungicide concentration. The coefficient `speciesB` tells us that the intercept of species B is around 67 units lower than the intercept of species A. Likewise the coefficient `speciesC` indicates that for this species the intercept is 45 units lower than for species A. In the model summary, the level of the categorical variable that appears first in the alphabet is used as a reference (baseline) level and called the intercept. The remaining factor levels are listed alphabetically and indicate the difference between their very own intercept and that of the reference level (the coefficient `speciesB` gives the difference between the intercept of species B and the intercept of species A, see annotated `summary(m1)` output). So, when we have a categorical predictor (factor) with two levels, then one is simply labelled `intercept`, and the other one is listed with its actual name in the summary output of the model.

Tip: The alphabetically determined intercept, serving as a reference level, can be changed by recoding the levels of the categorical variable, e.g.: `pathfung$species <- factor(pathfung$species, levels = c("B", "A", "C")) levels(pathfung$species)`

In this case, a significant difference between the two factor levels can be seen as an overall significance of this predictor. If we have more than two levels in a categorical predictor as in our example on fungal species, then the model summary first lists the reference level given by the intercept (representing the intercept of the alphabetically first factor level) followed by the differences between the respective 'factor level intercepts' and the intercept of the reference level.

However, in the first instance we often want to make a statement about the overall significance of a predictor, i.e. is there an overall statistically significant effect or not? In our case, we would be asking whether the fungicide has a significant effect on fungal growth (is the slope significantly different from zero) and whether there is a significant species effect (significantly different intercepts)? To obtain such overall significances, we can use the `drop1` command, which runs a model comparison procedure in the background, comparing the current model to a simpler model lacking (*dropping*) a predictor variable altogether. The model comparison relies on an F-test or a likelihood ratio test. A significant result lends support to the more complex model, i.e. the one including the explanatory variable in question. This means that a significant P-value associated with a model comparison can be interpreted as a significant effect of the dropped explanatory variable (which should therefore not be dropped from the final model).

```
## Use the drop1 command to perform an F-test based model comparison

> drop1(m1, test = "F")

Single term deletions

Model:
growth ~ conc + species

                                                        Sum                       F
                                                  Df  of Sq    RSS    AIC     value    Pr(>F)
Original model (no term dropped) ⬅ <none>                    21952 332.41
 Simpler model w/o concentration ⬅ conc      1 123568 145519 432.55 281.455  < 2e-16 ***
       Simpler model w/o species ⬅ species   2  41743  63694 385.93  47.539 2.72e-12 ***
---
Signif. codes:  0 '***' 0.001 '**' 0.01 '*' 0.05 '.' 0.1 ' ' 1
```

In our case, the `drop1` command ran two models in the background, one lacking fungicide concentration as a predictor (labelled `conc` in the output), and the other one lacking species (labelled `species`). These simpler models were then compared to the original model (labelled `<none>` short for 'none of the predictors dropped'). The `drop1` output tells us that both `conc` and `species` have statistically significant effects and should not be dropped from the model.

Our model `m1` contains only the main effects of fungicide concentration and species identity but no interaction term (fungicide concentration × species interaction). As indicated earlier, this means that the model assumes that the three fungal species share a common slope. Now let us see if this assumption is justified.

In theory, all factor levels (here all species) could show similar slopes, but this is often not the case in reality. In our example, it seems that a model allowing separate slopes for each fungal species could provide a better fit. We can achieve this by incorporating a fungicide concentration × species interaction in our ANCOVA model. The interaction between a continuous and a categorical predictor means that the slopes of the regression lines are allowed to vary with the levels of the categorical predictor variable. In our example, a statistically significant interaction would mean that the effect of the fungicide concentration on fungal growth rate varies across species.

```
## ANCOVA with interaction
> m2 <- lm(growth ~ conc * species, data = pathfung)

## Add model predictions to the data frame
> pathfung$fits2 <- predict(m2)

## Add the model predictions (fitted lines) to the plot
> lines(fits2 ~ conc, data = pathfung[pathfung$species == "A", ])
> lines(fits2 ~ conc, data = pathfung[pathfung$species == "B", ])
> lines(fits2 ~ conc, data = pathfung[pathfung$species == "C", ])
```

Adding the model predictions (regression lines) of the ANCOVA model with interaction to the plot nicely visualises the interaction: species A shows a much steeper slope than the two remaining species (Figure 11.1, right panel).

Let us move on to the `summary` output of our model `m2` including the fungicide concentration × species interaction, which is quite informative albeit somewhat harder to interpret. The explanations added to the output will help with the interpretation of the model parameters (= model coefficients).

```
> summary(m2)

Call:
lm(formula = growth ~ conc * species, data = pathfung)

Residuals:
    Min      1Q  Median      3Q     Max
-38.485  -7.030   0.773   6.346  45.982

                                                          Std.      t
                                  Coefficients  Estimate Error  value  Pr(>|t|)
      Intercept (Int.) speciesA ← (Intercept)    236.306 6.193  38.158  < 2e-16 ***
                 Slope speciesA ← conc           -19.759 1.023 -19.320  < 2e-16 ***
    Int. speciesB - Int. speciesA ← speciesB    -117.632 8.758 -13.431  < 2e-16 ***
    Int. speciesC - Int. speciesA ← speciesC     -80.241 8.758  -9.162 4.10e-12 ***
Slope speciesB - slope speciesA ← conc:speciesB   10.158 1.446   7.023 6.78e-09 ***
Slope speciesC - slope speciesA ← conc:speciesC    7.105 1.446   4.913 1.09e-05 ***
---
Signif. codes:  0 '***' 0.001 '**' 0.01 '*' 0.05 '.' 0.1 ' ' 1
```

As before, the `intercept` represents the intercept for species A but now the coefficient `conc` indicates the slope for species A only. The next two coefficients, `speciesB` and `speciesC`, indicate the deviation of the intercepts of species B and C from that of species A, respectively. Now comes the tricky bit: the coefficient `conc:speciesB` gives the difference in slopes between species B and species A. Because the linear relationship we see in our plot is decreasing for each of the fungal species, we expect negative slopes for all of them and it is thus clear that the positive values for `conc:speciesB` and `conc:speciesC` are not referring to the true slopes. The actual slope for species B is calculated by adding the estimate of the `conc:speciesB` interaction term to the slope of species A (`conc`). And the same approach is used to obtain the slope for species C:

$$slope_B = -19.759 + 10.158 = -9.601$$

$$slope_C = -19.759 + 7.105 = -12.654$$

As a plausibility check, we could simply use the intercept and slope for each of the fungal species to redraw the regression lines. This is basically what the predict function is doing. If our calculations are correct, the resulting lines should fall right onto the model predictions derived from the `predict` function (we leave this exercise to the keen reader who can plug the intercept (`a`) and slope (`b`) values into the function `abline(a = ... , b = ...)`).

The `drop1` command gives us an overall significance for the interaction term (performs an F-test comparing an ANCOVA with a concentration × species interaction to an ANCOVA without interaction).

```
> drop1(m2, test = "F")
Single term deletions

Model:
growth ~ conc * species
             Df Sum of Sq   RSS    AIC F value    Pr(>F)
<none>                    10543 296.81
conc:species  2     11409 21952 332.41   25.97 2.27e-08 ***
---
Signif. codes:  0 '***' 0.001 '**' 0.01 '*' 0.05 '.' 0.1 ' ' 1
```

Noting the significant interaction term, our job seems done (i.e. it is worth including the interaction term). However, before we use the model for inference or prediction, we need to check the underlying assumptions of normality and variance homogeneity of the residuals. As before, we assess these assumptions with the help of model diagnostic plots.

```
## Model diagnostic plots
> par(mfrow = c(2, 2))
> plot(m2)
```

The plot of the residuals vs. fitted values and the scale-location plot both indicate variance heterogeneity revealed by the fan-shaped pattern in the former and the increasing residual values with increasing fitted values in the latter (Fig. 11.2). The quantile–quantile plot suggests no gross deviation from normality (the deviation from the 1 : 1 line on the left can be considered mild) and the residuals vs leverage plot does not highlight any strongly influential observations (Figure 11.2, right-hand panels).

As we have seen before, variance heterogeneity (heteroscedasticity) is inherent in many types of scientific data. In fact, >90 % of the datasets we have been dealing with suffered from heteroscedasticity. From a statistical point of view though, variance heterogeneity is a problem that needs to be addressed. Unaccounted for variance ends up in the standard errors of the model coefficients, meaning that the variance of the standard errors is not constant across observations (incorrect standard errors). Usually, but not always, variance heterogeneity leads to an inflation of the standard errors, which in turn shrinks the t-statistic (which is simply the coefficient divided by its standard error), eventually resulting in inflated P-values. Put simply, if we ignore this most common type of heteroscedasticity, we are less likely to detect significant effects (increased probability of a type II error and thus reduced statistical power).

Because the interaction term in our example is overwhelmingly significant, the variance heterogeneity issue is not a major concern. However, in formal analyses, heteroscedasticity should always be dealt with and this can be readily achieved through variance modelling by

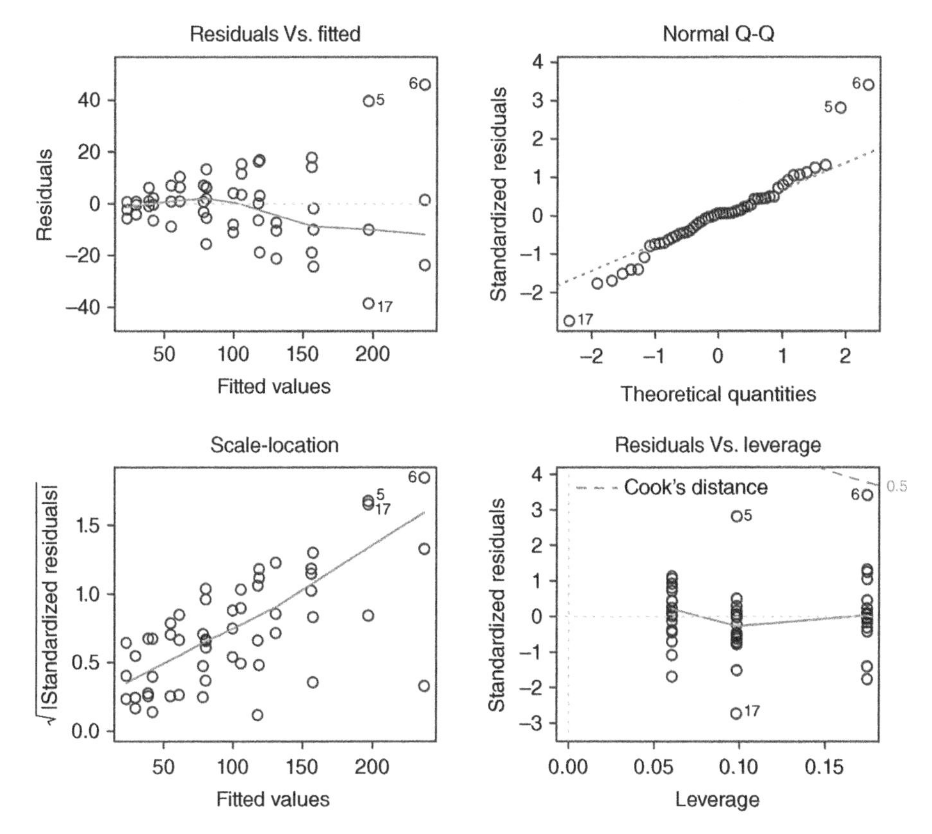

Figure 11.2 Diagnostic plots of the ANCOVA model with interaction (m2).

switching to a generalised least-squares model (gls, R package *nlme*, Pinheiro et al. 2022), which allows the incorporation of variance functions. These variance functions in gls models can be thought of as small submodels that describe the heteroscedasticity pattern in the data and are used to assign weights to the observations, basically resulting in a fancy sort of weighted linear regression. This elegant approach goes beyond the scope of this textbook (see Section 12.4), but we refer to the excellent texts of Zuur et al. (2009) and Pinheiro & Bates (2000).

In the presence of a significant interaction term, the main effects of fungicide concentration and fungal species are not properly interpretable and thus need not be elaborated on (Zuur et al. 2009). Some scientific papers may still report the *P*-values of the main effects along with that of a significant interaction, but following the reasoning by Zuur et al. (2009), we advise against this and suggest to focus on the interaction.

In a paper, thesis or report, one would state the overall significance of the interaction term along the lines of: 'The efficacy of the tested fungicide concentrations varied with fungal species as evidenced by a significant fungicide × species interaction ($F_{2,48} = 25.97, P < 0.001$).'

Recall that the *F*-statistic follows an *F*-distribution whose shape depends on the numerator and denominator degrees of freedom (see Chapter 10). The numerator and denominator degrees of freedom are usually given as the subscripts after the *F*-value. How do we get those again?

The numerator degrees of freedom are listed in the drop1 output (number of factor levels minus one). The denominator degrees of freedom (also called residual degrees of freedom) can be obtained using the command `df.residual` (they are also listed at the end of the `summary(m2)` output). These are simply the number of observations in the data frame minus the number of model coefficients, which we can check with the following code:

```
## Residual degrees of freedom
> df.residual(m2)
[1] 48

## Residual degrees of freedom calculated by hand
> nrow(pathfung) - length(coef(m2))
[1] 48
```

If the interaction had not been significant, we could have safely removed it to simplify the model.

11.2 *Post Hoc* Test for ANCOVA

Now that we have found out that there is indeed a significant fungicide concentration × species interaction, the question is which slopes differ significantly from each other? A significant interaction term could result from a significant difference between only two of the slopes but such differences could just as well occur between all slopes and this is something we follow up on with a *post hoc* analysis. Often, we are interested in all pairwise comparisons (comparing all slopes with each other) and we can perform this procedure automatically including a *P*-value adjustment to correct for the increase in **Type I error** probability involved with multiple comparisons on the same dataset. The *emmeans* package provides the `emtrends` function, which returns individual slope estimates and associated confidence intervals and allows pairwise comparisons between slopes (Lenth 2022).

```
## Multiple comparisons for an ANCOVA model with interaction
## Obtain slope estimates incl. confidence intervals for each species
> slopes <- emtrends(m2, specs = "species", var = "conc")
> slopes
```

Species	Conc.trend	SE	df	Lower.CL	Upper.CL
A	−19.8	1.02	48	−21.8	−17.70
B	−9.6	1.02	48	−11.7	−7.55
C	−12.7	1.02	48	−14.7	−10.60

```
## Apply a pair-wise comparison procedure to the slopes
> pairs(slopes)
```

Contrast	Estimate	SE	df	t.ratio	p.value
A − B	−10.168	1.45	48	−7.023	<.0001
A − C	−7.11	1.45	48	−4.913	<.0001
B − C	3.05	1.45	48	2.110	0.0983

```
P value adjustment: tukey method for comparing a family of 3 estimates
```

Pairwise comparisons, which form the basis of our *post hoc* analysis, rely on testing whether the difference between two slopes is significantly different from zero (obviously a difference of zero or close to zero suggests similar slopes). With this knowledge, it becomes easy to read the output of the *post hoc* analysis: the ‘contrast’ column indicates the two species whose slopes were compared. The ‘estimate’ column gives the differences between the two slopes.

The 'SE' column indicates the standard error around the estimated differences in slopes. The residual degrees of freedom are listed in the 'df' column. The values in the 't.ratio' column represent the *t*-statistic (estimate divided by SE) and the 'p.value' column contains the associated *P*-values.

We can interpret the *post hoc* analysis results as follows: Species A shows a significantly steeper slope, i.e. a stronger growth decline in response to increasing fungicide concentration, than the remaining two species whose slopes do not differ significantly from each other.

References

Lenth, R.V. (2022). emmeans: Estimated marginal means, aka least-squares means. [R package Version 1.8.3] (2022). R-project.org. [online]. https://CRAN.R-project.org/package=emmeans.

Pinheiro, J.C. and Bates, D.M. (2000). *Mixed-effects models in S and S-PLUS*. New York: Springer.

Pinheiro, J., Bates, D. and R Core Team Linear and Nonlinear Mixed Effects Models [R package nlme version 3.1-141]. (2022). R-project.org. [online]. https://CRAN.R-project.org/package=nlme.

Zuur, A.F., Ieno, E.N., Walker, N.J. et al. (2009). *Mixed Effects Models and Extensions in Ecology with R*. New York: Springer.

12

Some of What Lies Ahead

In this final chapter, we provide a teasing outlook on some important, widely used data analysis techniques and remedies for heteroscedasticity, whose in-depth treatment would go beyond the scope of this introductory text. Therefore, we ask the readers to bear in mind that the illustrative presentation of these techniques inevitably falls short of being exhaustive. In Box 12.1, we summarise and exemplify the use of various important *tidyverse* functions (R package *tidyverse*, Wickham et al. 2019), some of which we have already seen in action in previous chapters. In addition, we provide a dichotomous key to the linear models covered in this textbook but we also go a bit further and point to extensions related to frequently encountered types of data, such as counts or rating scales, to give the reader a broader perspective and to avoid flawed analysis approaches (Box 12.2).

12.1 Generalised Linear Models

We will start with *generalised linear models* (GLMs), which represent a versatile generalisation of the linear model. GLMs relax the assumption of normally distributed errors by allowing alternative error distributions, which makes it possible to analyse discrete data (response variables) such as counts, binary data, and continuous variables with non-normal errors (e.g. right-skewed data characterised by a few extremely high values).

GLMs have three basic components:

(1) a **distribution** for the response variable and the model errors.
(2) a **linear predictor** that wraps up all the regression parameters and covariates.
(3) a **link function** that transforms the predictor function to the scale of the response variable (links the linear predictor to the expected values of the response variable).

Use GLMs to analyse discrete response variables.

We fit GLMs with a function called `glm`, which uses a similar syntax to the `lm` function. The main difference is that we need to specify a distribution family for the response variable/model errors (see help for `family` to view the available distributions and their possible link functions). Note that you can choose between multiple link functions for a certain distribution. When using the default link, it is sufficient to specify the family only, i.e. `family = binomial(link = "logit")` is equivalent to `family = binomial`.

We will exemplify the use of GLMs using binary data, which is commonly referred to as *logistic regression*. A binary response variable may only take on two possible values such as dead/alive, absent/present, male/female, parasitised/unparasitised, etc. Binary data follow a binomial distribution which describes the discrete probability distribution of the number

R-ticulate: A Beginner's Guide to Data Analysis for Natural Scientists, First Edition.
Martin Bader and Sebastian Leuzinger.

Companion website: www.wiley.com/go/Bader

of successes in a sequence of success/failure trials (see Chapter 3). If we examine say 100 toads in a habitat for the occurrence of parasites, then this can be viewed as 100 individual success/failure trials, just like a binary series of heads or tails in the oft-cited coin tossing example. It is left to the researcher to decide on how 'success' and 'failure' are defined. In this instance, where the research question revolves around parasitism, one would perhaps regard a parasitised individual as a 'success' and record such an observation as a '1' and unparasitised specimens would be considered 'failure' and recorded as a '0'. The binary response can be specified in a GLM in three different ways:

(1) as a two-level factor such as yes/no, alive/dead, which will be treated as a 0/1 vector (or simply provided as a vector of 0s and 1s from the start).
(2) as a proportion, in which case the number of trials (observations) must be provided as a vector of weights using the `weights` argument in the `glm` function.
(3) as a two-column matrix holding the number of *successes* in the first and the number of *failures* in the second column.

The last option is perhaps the least intuitive way but provides better starting values for the fitting algorithm and should therefore be the preferred method. As an example, we will use a dataset of pesticide efficacy on a bark beetle species. For each tested pesticide concentration, batches of 20 beetles were exposed over 24 hours. The response variable was the number of dead individuals recorded in 0/1 format (1 = dead, 0 = alive). A contingency table provides a more informative overview of binary data than `str` or `summary`.

```
## Binomial model for pesticide efficacy data
## Load required packages
> library(ggplot2)
> library(dplyr)

> ## Read in data
> pest <- read.csv("pesticide.csv")
> addmargins(table(pest)) # contingency table (frequency table)
      dead
dose    0   1 Sum
  0    19   1  20
  2    17   3  20
  4    13   7  20
  8     5  15  20
  10    2  18  20
  15    0  20  20
  Sum  56  64 120
```

The contingency table shows that the pesticide efficacy increases with increasing dose, culminating in all 20 beetles dead at the highest dose.

It is not very enlightening to plot 0s and 1s (try it yourself!). Often, as in the present case, we can express binary data as proportions, resulting in a much easier-to-interpret plot (and multiplying proportions by 100 yields percentages, Figure 12.1a, but see also the use of `mosaicplot` in Chapter 7). In fact, in logistic regression, the predicted values are probabilities giving the likelihood of the binary outcome. For simplicity and ease of interpretation, you can view these probabilities as proportions.

```
## Calculate the proportion of dead beetles for each pesticide dose
> pest2 <- group_by(pest, dose) %>%
summarise(dead = sum(dead), n = n(), prop = dead/n)) # n() counts the obser-
vations per dose (here 20)

## Plot the proportions
> ggplot(data = pest2, aes(x = dose, y = prop)) +
```

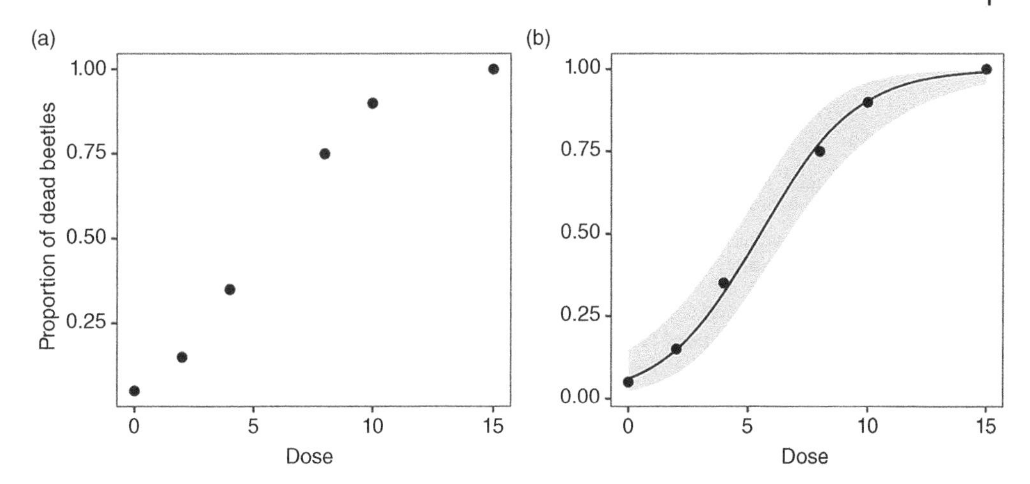

Figure 12.1 (a) Binary data (dead/alive) displayed as proportion of dead beetles at each pesticide dose. (b) Binomial GLM fit (solid black line) and surrounding 95% confidence interval (grey area) added to the observed proportions of dead beetles. Please note how the lower and upper limits of the confidence interval correctly taper off as they approach 0 or 1, respectively, which is due to the logit transformation that prevents the occurrence of nonsensical out-of-range values (it is not possible to have a probability less than 0 or greater than 1).

```
    labs(x = "Dose", y = "Proportion of dead beetles") +
    geom_point() +
    theme_test()
```

For the sake of completeness, we present all three ways of fitting a GLM with binomial error distribution. We will stick with the default logit link, which ensures that the predicted probabilities stay within the 0–1 range.

```
## GLM specification using a two-column matrix of success (dead beetles)
and failure (live beetles) as response variable
## Create the two-column success/failure matrix
> y <- with(pest2, cbind(success = dead, failure = n - dead)) # 'with'
allows us to access variables without the need to explicitly refer to the
data frame

> y
     success failure
[1,]       1      19
[2,]       3      17
[3,]       7      13
[4,]      15       5
[5,]      18       2
[6,]      20       0

## Fit the logistic regression model (a.k.a. binomial GLM)
> glm1 <- glm(y ~ dose, data = pest2, family = binomial()) # binomial()
defaults to binomial(link = "logit")
> summary(glm1)

Call:
glm(formula = y ~ dose, family = binomial(link = "logit"), data = pest2)

Coefficients:
            Estimate Std. Error z value Pr(>|z|)
(Intercept) -2.75102    0.51261  -5.367 8.02e-08 ***
dose         0.49953    0.08271   6.039 1.55e-09 ***
—
Signif. codes:  0 '***' 0.001 '**' 0.01 '*' 0.05 '.' 0.1 ' ' 1
```

```
(Dispersion parameter for binomial family taken to be 1)

    Null deviance: 79.57803  on 5  degrees of freedom
Residual deviance:  0.54876  on 4  degrees of freedom
AIC: 18.415

Number of Fisher Scoring iterations: 4

## Alternative model specification using the 0/1 data in the original dataset
# > glm2 <- glm(dead ~ dose, data = pest, family = binomial())
# > summary(glm2)

## Alternative model specification using proportion data with the num-
ber of trials (n = 20 beetles per dose) supplied by the weights argument
# > glm3 <- glm(prop ~ dose, data = pest2, family = binomial(), weights = n)
# > summary(glm3)
```

Fitting a standard linear regression model would produce unrealistic predictions for the binary response variable (below 0 and above 1).

Now we calculate the model predictions and their standard errors to construct a 95% confidence interval. It is important to note that *confidence intervals or upper and lower bounds of error bars* (mean ± standard error) *must be calculated on the link scale*, i.e. prior to back-transforming from the logit scale to probabilities. This is necessary to make sure that the predicted values including their measures of uncertainty do not fall outside the permissible range between 0 and 1 to which probabilities are limited.

```
## Set up a fine grid of equally spaced new dose values to predict from
(ensures a smooth fitted curve)
> newdat <- data.frame(dose = seq(min(pest2$dose), max(pest2$dose),
length.out = 200))
```

All distribution families have a 'hidden' function built-in to facilitate back-transformation from the link scale to the response scale (see `linkinv` in the 'Value' section of the help file for `family`).

```
## Get model predictions and their standard errors on the link scale
(because the 95 % CI needs to be calculated on the link scale)
> preds <- predict(glm1, newdata = newdat, se.fit = T)

## Add the back-transformed GLM fit and the 95 % confidence interval to the
newdat data frame
> bt <- binomial()$linkinv # custom function to back-transform predictions
from the logit scale to the response scale (a probability)

## IMPORTANT: The confidence interval must be calculated on the link scale,
i.e. before back-transformation to the response scale.
> newdat$fit <- bt(preds$fit)
> newdat$upper <- bt(preds$fit + 1.96 * preds$se.fit)
> newdat$lower <- bt(preds$fit - 1.96 * preds$se.fit)
```

As a consequence of the logit transformation, the confidence interval turns out asymmetrical, which is a seal of quality for a correct modelling procedure rather than a flaw. Since the range of the probabilities is bounded by 0 and 1, the associated confidence interval is limited in the same way and the logit transformation (logit link) ensures that the model predictions and their confidence intervals do not spill over these boundaries. Therefore, the upper and lower confidence limits must inevitably taper off towards 1 and 0. *Symmetrical confidence intervals* of a logistic regression fit exceeding 0 or 1 indicate an *incorrect calculation* that was performed *after back-transformation* from the link scale.

Asymmetric confidence intervals are characteristic of GLMs!

Note that despite the nonlinear appearance of the logistic regression curve, the relationship between the predictor(s) and the binary response variable is linear on the logit scale. The nonlinear appearance is a result of the back-transformation from the logit scale to the response scale, i.e. the probability of a binary event occurring.

We have skipped the model validation step here, but in practice one would first inspect the model diagnostic plots using the `glm.diag` and `glm.diag.plots` functions in R package *boot* (see the help for `glm.diag.plots` for an example, Canty and Ripley 2022; Davison and Hinkley 1997). Various pseudo-R^2 measures for GLMs are available via the `pR2` function in R package *pscl* (Jackman 2020).

Box 12.1

Concise overview of some important *tidyverse* functions. *Tidyverse* is a collection of R packages providing powerful tools for data manipulation and visualisation. The nine core packages forming *tidyverse* are: *ggplot2*, *dplyr*, *tidyr*, *readr*, *purrr*, *tibble*, *stringr*, *lubridate*, and *forcats*. Here we focus on the *dplyr* and *tidyr* packages. Multiple *ggplot2* examples can be found throughout this textbook. More information on the *tidyverse* packages including cheat sheets can be found here https://www.tidyverse.org/packages/

You will need to load the built-in dataset `airquality` by typing `data(airquality)` to run the examples (see help file for this dataset for more information). Note that the pipe operator `%>%` does not replace the assignment operator `<-`, so storing pipe operations as a new object requires an assignment.

Function/ operator	R package	Description	Examples

`%>%`
`|>`
	magrittr (automatically re-imported by various packages)
	base
		The pipe operator forwards an object to a function. This allows chaining operations together in a pipeline-like fashion, which makes the code more readable and closer to natural language.
		In RStudio, the keyboard shortcut for the pipe operator is Cmd + Shift + M (Mac) and Ctrl + Shift + M (Windows)
		See Wickham (2023) for differences between the *base* R and *magrittr* pipes
			`# Order dataset in ascending fashion by temperature`
			`airquality %>% arrange(Temp)`
			`# Same in descending fashion`
			`airquality %>% arrange(desc(Temp))`
			`# Subset data by observations with ozone concentration greater than 100 ppb`
			`airquality %>% filter(Ozone > 100)`
			`# Subset only containing May observations`
			`airquality %>% filter(Month == 5)`
			`# Create a new variable giving a wind dispersion index`
			`airquality %>% mutate(dispersion = Ozone/Wind)`
			`# Note that in all of the above examples, you need to create a new object using the assignment operator <- to store your work, e.g.:`
			`airquality2 <- airquality %>% mutate(dispersion = Ozone/Wind)`

(Continued)

Box 12.1 (Continued)

Function/ operator	R package	Description	Examples
`group_by`	*dplyr*	Groups a dataset by one or more variables to allow group-wise data operations.	See below.
`summarise` `summarize`	*dplyr*	Summarises each group in a dataset (or the entire dataset) down to one row.	`# Mean and standard deviation of air temperature by month` `airquality %>% group_by(Month) %>% summarise(tmean = mean(Temp), sd_temp = sd(Temp))` `# Means for all variables (apart from day) by month` `airquality %>% group_by(Month) %>% summarise(across(.cols = -last_col(), .fns = function(x) mean(x, na.rm = T)))`
`reframe`	*dplyr*	Similar to `summarise` but instead of only one row (per group), `reframe` can return an arbitrary number of rows, which allows the use of more complex functions that provide more than a single summary value.	`# Compute quantiles of the ozone concentration by month` `airquality %>% group_by(Month) %>% reframe(ozone = quantile(Ozone, na.rm = T)) %>% mutate(quant = seq(0, 1, by = 0.25), .by = Month)`
`pivot_wider`	*tidyr*	Pivot data from long to wide format, i.e. increasing the number of columns and reducing the number of rows.	`# Create individual columns for each month and assign temperatures` `airq_wide <- airquality %>% select(Temp, Month, Day) %>% group_by(Month) %>% pivot_wider(names_from = Month, values_from = Temp)`
`pivot_longer`	*tidyr*	Pivot data from wide to long format, i.e. increasing the number of rows and decreasing the number of columns	`# Convert the data in wide format into a long data frame with a month and a temperature variable` `airq_long <- airq_wide %>% pivot_longer(cols = 2:6, names_to = "month", values_to = "temp")`

Function/ operator	R package	Description	Examples
`unite`	*tidyr*	Combine multiple columns into one by pasting strings together	`# Combine month and day into a date variable` `airq <- airquality %>% unite(col = "date", c(Month, Day), sep = "-")`
`separate_wider_delim`	*tidyr*	Split a character string into separate columns	`# Split the date variable into month and day` `airq %>% separate_wider_delim(cols = date, delim = "-", names = c("month", "day"))`

12.2 Nonlinear Regression

Nonlinear equations are frequently of mechanistic nature (based on a theoretical understanding of the system at hand). The model parameters therefore often have a meaning regarding the underlying processes.

Many processes in nature are nonlinear and can be modelled using nonlinear regression. The underlying mathematical equations often represent biological processes so that the parameters have a biological meaning, i.e. the models have a mechanistic character. To perform a nonlinear regression in R, we use the function `nls` which stands for ***nonlinear least-squares*** regression, which is based on minimising the sum of the squared residuals (residual sum of squares), just like the ordinary least squares method for linear regression.

We will use the relationship of photosynthesis and light intensity as an example. With increasing light intensity (predictor variable), the rate of photosynthesis (response variable) increases linearly as long as light is the main limiting factor, but levels off into a plateau as the dark reactions of the photosynthetic process become limiting (Calvin cycle) (Figure 12.2a).

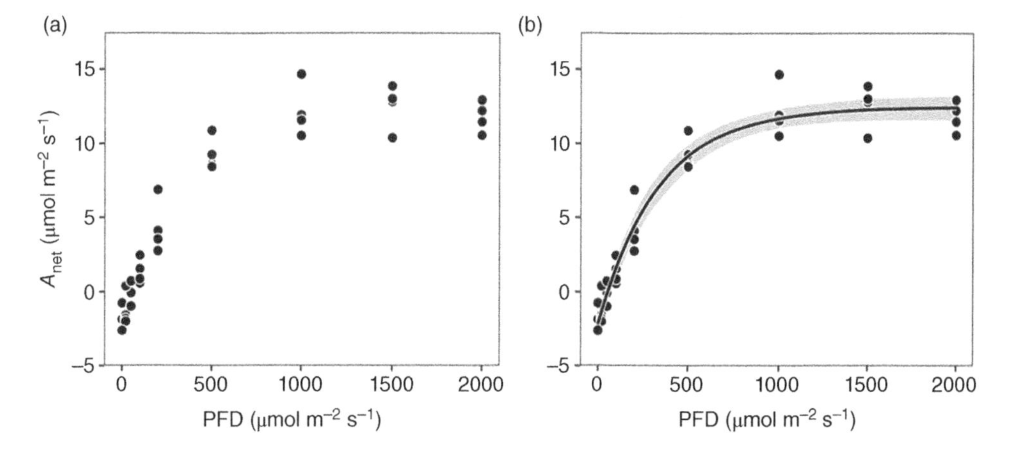

Figure 12.2 (a) Leaf photosynthesis as a function of light intensity (photon flux density, PFD). (b) Nonlinear regression curve (solid black line) and associated 95% confidence interval (grey area) fitted to the data in (a).

This relationship can be described by various models (exponential, rectangular hyberbola, and non-rectangular hyperbola) and here we will use the following exponential model:

$$A_{\text{net}} = A_{\max}\left(1 - e^{-\alpha\, PFD/A_{\max}}\right) + R_d$$

where the response variable A_{net} is the net photosynthetic rate (μmol CO_2 m^{-2} s^{-1}), the predictor variable PFD is the photon flux density (light intensity) (μmol quanta m^{-2} s^{-1}), and $A_{\max}$, α, and R_{d} are model parameters estimated by the nonlinear least-squares algorithm. $A_{\max}$ is the maximum photosynthesis rate, α is the quantum yield (initial slope of the linear section of the curve, a dimensionless quantity), and R_{d} is the dark respiration rate (μmol CO_2 m^{-2} s^{-1}), i.e. the respiratory CO_2 released from leaves measured in the absence of light.

We first read in the data and conduct a plausibility check.

```
## Install and load required packages
> install.packages("nlraa")
> library(nlraa)
> library(ggplot2)

## Read in the data
> lc <- read.csv("photosynthesis_light_response.csv", header = T) # 'lc'
stands for 'light curve'

> str(lc)
'data.frame':       36 obs. of  3 variables:
 $ photo      : num  -1.8455 -1.5703 -0.0586 1.5415 4.1175 ...
 $ pfd        : int  0 20 50 100 200 500 1000 1500 2000 0 ...
 $ individual: int  1 1 1 1 1 1 1 1 1 2 ...

> summary(lc)
     photo               pfd               individual
 Min.    :-2.6732   Min.    :    0.0    Min.    :1.00
 1st Qu.:-0.2315   1st Qu.:   50.0    1st Qu.:1.75
 Median : 3.8305   Median : 200.0    Median :2.50
 Mean    : 5.3336   Mean    : 596.7    Mean    :2.50
 3rd Qu.:11.0380   3rd Qu.:1000.0    3rd Qu.:3.25
 Max.    :14.6699   Max.    :2000.0    Max.    :4.00
```

We can see that photosynthesis (`photo`) ranges from negative values around −2.7 μmol m^{-2} s^{-1}, indicating leaf respiration in the dark (CO_2 release), to maximal photosynthesis rates close to 15 μmol m^{-2} s^{-1} (CO_2 uptake). Light intensities (`pfd`) range from zero (darkness) to 2000 μmol CO_2 m^{-2} s^{-1} equivalent to full sunlight. The `individual` variable indicates that the experiment was replicated with four individual plants. As the `str` and `summary` output do not show any obvious issues (such as unrealistically low or high values), we can go ahead with plotting the data to get a visual impression of the photosynthesis–light relationship (Figure 12.2a).

```
## Pre-define axis titles
> ylab <- expression(paste(italic(A)[net], " (", mu, "mol m"^-2,
" s"^-1,")"))
> xlab <- expression(paste("PFD (", mu, "mol m"^-2," s"^-1,")"))

## Plot of the raw data
> ggplot(data = lc, aes(x = pfd, y = photo)) +
  geom_point() +
  labs(x = xlab, y = ylab) +
  theme_test()
```

12.2.1 Initial Parameter Estimates (Starting Values)

In contrast to linear regression, where the intercept and the slope can be readily estimated by the model algorithm, we need to provide initial parameter estimates, the so-called starting values, to get the nonlinear model algorithm going. This can be a daunting task if the parameters do not have an obvious meaning or when they lie within a very narrow range where minute changes have a great effect. In our photosynthetic light curve example, however, we can readily deduce from the underlying equation that the two parameters A_{max} and R_d may be estimated by the maximum and minimum photosynthesis values in our dataset.

Estimating the quantum yield (α, the intitial slope of the curve) is slightly more elaborate. The slope of a linear regression model applied over the first, say three or four light levels, commonly yields a fairly good estimate of α. We can do this on the fly by running a linear model on a truncated dataset confined to low light intensities.

Nonlinear least-squares models require starting values (initial guesses) for parameter estimation.

```
## Deriving initial parameter estimates to plug into the nonlinear model
> max(lc$photo) # initial estimate for Amax
> min(lc$photo) # initial estimate for Rd


## Linear model for the first 4 light intensities, i.e. the quasi-linear
part of the response pattern. The slope gives an estimate of alpha.

> lm1 <- lm(photo ~ pfd, data = lc[lc$pfd <= 100, ]) # <= means smaller than
or equal to

> coef(lm1) # extracts the model coefficients
(Intercept)         pfd
-1.94143523  0.03305027

> coef(lm1)[2] # extracts the slope as an estimate for the alpha parameter
[1] 0.03305027
```

Self-starter functions provide a routine for calculating starting values for a specific nonlinear regression equation. That means the user does not have to worry about initial parameter values making nonlinear regression modelling a painless and enjoyable experience. The base installation of R (see at the bottom of the help file for the `selfstart` self function under the section titled 'See Also') and packages *drc* (Ritz et al. 2015), *nlraa* (Miguez 2023), and *aomisc* (only available on Github, Onofri 2020) come with a collection of built-in self-starter functions. With some practice you can also create self-starter functions yourself. The excellent textbook on nonlinear regression by Ritz and Streibig (2008) provides instructions and a worked example.

12.2.2 Nonlinear Model Fitting and Visualisation

We can now plug the parameter estimates into the `nls` function via the `start` argument. Note that they must be provided as a named list (`list` function) or vector (`c` function), matching the names in the nonlinear equation.

```
## Nonlinear model
> nlm1 <- nls(photo ~ Amax * (1 - exp(-alpha * pfd/Amax)) + Rd, data = lc,
              start = list(Amax = max(lc$photo),
```

```
                                    Rd = min(lc$photo),
                                    alpha = coef(lm1)[2]))

> summary(nlm1)

Formula: photo ~ Amax * (1 - exp(-alpha * pfd/Amax)) + Rd

Parameters:
       Estimate Std. Error t value Pr(>|t|)
Amax  14.650803   0.499045  29.358  < 2e-16 ***
Rd    -2.197885   0.386753  -5.683 2.46e-06 ***
alpha  0.044131   0.005028   8.778 3.81e-10 ***
---
Signif. codes:  0 '***' 0.001 '**' 0.01 '*' 0.05 '.' 0.1 ' ' 1

Residual standard error: 1.178 on 33 degrees of freedom

Number of iterations to convergence: 5
Achieved convergence tolerance: 1.441e-06
```

The summary output indicates that all three parameter estimates are statistically significant (different from zero). Normally, we would inspect the diagnostic plots to assess the underlying assumptions first before using the model for inference, but for the sake of brevity, we omit this step and move straight on to the model predictions. The default `predict` function does not provide standard errors or confidence intervals. Therefore, we will use the `predict_nls` function, which allows the calculation of confidence intervals along with the mean nonlinear least-squares fit (Figure 12.2, R package *nlraa*, Miguez 2023).

```
## Get model predictions incl. 95 % confidence interval (95 % CI)
## Create a fine grid of new x-values (new predictor values to predict from)
> xv <- with(lc, seq(min(pfd), max(pfd), length.out = 100)) # xv = new
x-values (predictor values)

> newdat <- data.frame(pfd = xv) # Note that the predictor name (pfd) must
exactly match the predictor name in the nls function!

## Use predict_nls() from the nlraa package to obtain fits and 95 % CI
> preds <- as.data.frame(predict_nls(nlm1, newdata = newdat, interval =
"confidence"))

## Combine the new predictor values with the predictions and 95 % CI
> newdat <- cbind(newdat, preds)
> newdat$photo <- 0 # ggplot requires a vacuous photo variable as this
occurs in the original lc data frame (see ggplot code below)

> head(newdat, n = 3) # 'Estimate' indicates the model fit, Q2.5 and Q97.5
are the lower and upper confidence limits

       pfd   Estimate Est.Error       Q2.5       Q97.5 photo
1  0.00000 -2.1910922 0.3966916 -2.9781321 -1.44434503     0
2 20.20202 -1.3298816 0.3310253 -1.9942803 -0.71184521     0
3 40.40404 -0.5228718 0.2865679 -1.0996332  0.03361359     0


## Add the nonlinear regression fit and 95 % CI to the data

> ggplot(data = lc, aes(x = pfd, y = photo)) +
  geom_point() +
```

```
  labs(x = xlab, y = ylab) +
  geom_ribbon(data = newdat, aes(x = pfd, ymin = Q2.5, ymax = Q97.5), fill
= "grey80", colour = "grey80") +
  geom_line(data = newdat, aes(x = pfd, y = Estimate)) +
  theme_test()
```

Various helper functions, nonlinear equations, self-starter versions thereof and data examples are available in R packages *nlme* (Pinheiro et al. 2023), *nlraa* (Miguez 2023), *nlstools* (Baty et al. 2015), and *nlsMicrobio* (Baty and Delignette-Muller 2014). A rich compilation of nonlinear model equations can be found in Sit and Poulin-Costello's (1994) curve fitting handbook, which can be downloaded for free (https://www.for.gov.bc.ca/hfd/pubs/docs/bio/bio04.htm).

12.3 Generalised Additive Models

One could say that generalised additive models (GAMs) are a sort of new wonder weapon in data analysis as they can model any nonlinear trend. This makes them great exploratory tools for detecting nonlinear trends in data but also very powerful, full-fledged models that provide predictions and associated standard errors and allow model comparisons using standard tools (e.g. AIC or likelihood ratio tests).

GAMs are extensions of GLMs in which the linear predictor is given by a user-specified sum of smooth functions of the continuous predictor(s) (nonparametric part) plus a conventional parametric component linked to the categorical explanatory variables. In plain English, you could describe a GAM as a model that stitches together multiple smoothing splines to provide a highly flexible fit allowing us to capture intricate patterns that may not be well-described by linear or nonlinear regression models (if present).

Ok, this may sound quite technical but once you start applying GAMs to data, you quickly get the hang of it. The most important concept to grasp here is that GAMs use smoothing functions, usually based on splines, to capture nonlinear relationships between continuous predictors and the response variable(s). Just as with GLMs, we can choose different error distributions allowing us to model all sorts of response variables (Wood 2011; Wood 2017). In addition to the distributions that come with the base installation of R (see help for `family`), more distributions for GAMs are available in the *mgcv* package (see help for `family.mgcv`).

GAMs can capture ANY nonlinear pattern but do not represent the underlying processes (non-mechanistic)!

In some species of flowering plants, the flower diameter shows a bimodal distribution in adaptation to differently sized pollinators (Figure 12.3a). We will use such a dataset to illustrate how to fit a simple GAM.

```
## Install and load required package
> install.packages("mgcv")
> library(mgcv)

## Read in data
> flowers <- read.csv("flowers.csv")

> str(flowers)
'data.frame':       100 obs. of  2 variables:
 $ dia               : num  12.6 17.2 13 20.6 10.6 ...
 $ pollinator_length: num  7.77 15.97 11.81 15.32 20.11 ...
```

```
> summary(flowers)
      dia          pollinator_length
 Min.   : 5.00    Min.   : 5.00
 1st Qu.:10.73    1st Qu.:11.91
 Median :13.11    Median :17.87
 Mean   :13.98    Mean   :17.89
 3rd Qu.:17.16    3rd Qu.:23.83
 Max.   :25.00    Max.   :30.00
```

The data checking tools do not flag any issues, so we will plot the data to get a visual impression of the relationship.

```
> ggplot(data = flowers, aes(x = pollinator_length, y = dia)) +
  geom_point() +
  labs(y = "Flower diameter (mm)", x = "Pollinator length (mm)") +
  theme_test()
```

The plot clearly shows the bimodal relationship, which we will model with a GAM (function gam in R package *mgcv*, Wood 2011) assuming normally distributed flower diameters and hence model errors (Figure 12.3a). The default gam settings assume a Gaussian (normal) distribution, so in our case, there is no need to specify this explicitly. The gam function uses the familiar formula notation but the continuous predictor(s) are wrapped in a smoothing function that fits a regression spline.

```
## Fitting the GAM
> g1 <- gam(dia ~ s(pollinator_length), data = flowers)

> summary(g1)

Family: gaussian
Link function: identity

Formula:
dia ~ s(pollinator_length)

Parametric coefficients:
            Estimate Std. Error t value Pr(>|t|)
(Intercept)  13.9775     0.2672   52.32   <2e-16 ***
---
Signif. codes:  0 '***' 0.001 '**' 0.01 '*' 0.05 '.' 0.1 ' ' 1

Approximate significance of smooth terms:
                       edf Ref.df     F p-value
s(pollinator_length) 7.236  8.257 18.39  <2e-16 ***
---
Signif. codes:  0 '***' 0.001 '**' 0.01 '*' 0.05 '.' 0.1 ' ' 1

R-sq.(adj) =  0.604   Deviance explained = 63.3%
GCV =  7.779  Scale est. = 7.1384    n = 100
```

The model summary is two-parted, showing the *parametric component* at the top and the *nonparametric (smoother-related) component* at the bottom. In this simple example, the parametric part only comprises the intercept, which indicates 'how high the GAM fit is hovering' in relation to the *y*-axis.

The nonparametric part gives the significance of the smooth term(s). The important bits here are the estimated degrees of freedom (edf), which commonly pop up as decimals when smoothing is involved, the test statistic (*F*-value) and the resulting *P*-value indicating

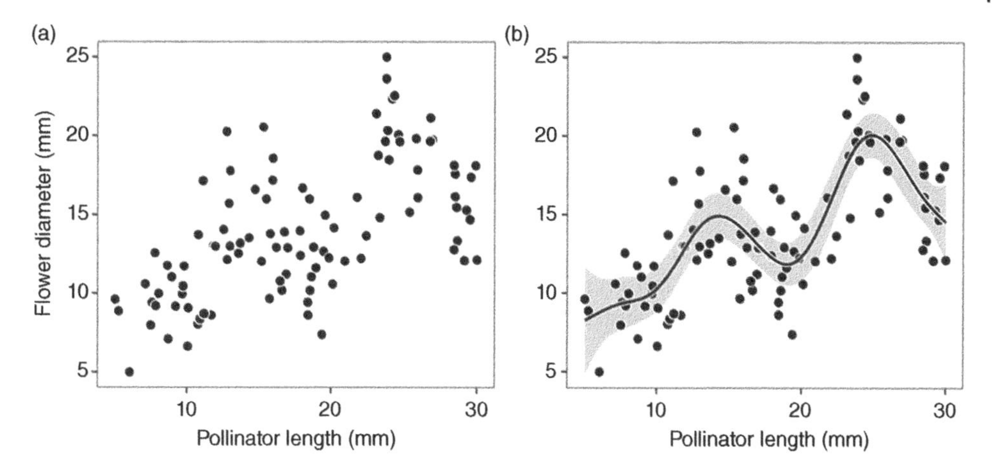

Figure 12.3 (a) Flower diameter in relation to pollinator length. (b) Generalised additive model fit (solid black line) and associated 95% confidence interval (grey area) fitted to the flower data.

whether the smoothing function is significantly different from zero. The larger the edf, the higher the degree of nonlinearity (edf = 1 indicates a linear relationship). In addition, two measures of goodness-of-fit are given, the adjusted R^2 (suitable for normally distributed errors, i.e. gaussian distribution) and the deviance explained (appropriate for non-normal errors, i.e. non-gaussian distributions). Getting the GAM predictions and the associated 95% confidence interval follows the standard procedure seen earlier:

```
> newdat <- data.frame(pollinator_length =
seq(min(flowers$pollinator_length), max(flowers$pollinator_length),
length.out = 100))
> preds <- predict(g1, newdata = newdat, se.fit = T)
> newdat$fit <- preds$fit
> newdat$upper <- preds$fit + 1.96 * preds$se.fit
> newdat$lower <- preds$fit - 1.96 * preds$se.fit
> newdat$dia <- 0 # required to satisfy ggplot expectations

> ggplot(data = flowers, aes(x = pollinator_length, y = dia)) +
  geom_ribbon(data = newdat, aes(x = pollinator_length, ymin = lower,
                ymax = upper), fill = "grey80", colour = "grey80") +
  geom_point() +
  geom_line(data = newdat, aes(x = pollinator_length, y = fit)) +
  labs(y = "Flower diameter (mm)", x = " Pollinator length (mm)") +
  theme_test()
```

The GAM fit neatly captures the bimodal pattern in the data, allowing us to separate the differently adapted subpopulations and to report their mean flower diameters including uncertainty measures such as standard errors or confidence intervals (Figure 12.3b). Again, for brevity reasons we have skipped model validation, which can be done using the `gam.check` function (R package *mgcv*, Wood 2011).

12.4 Modern Approaches to Dealing with Heteroscedasticity

The homoscedasticity (variance homogeneity) assumption of linear models is often violated, which usually has little, if any, influence on the parameter estimates but typically inflates or reduces the associated standard errors and thus confidence intervals, resulting in biased inference. We have previously pointed out that there are more modern and sophisticated tools than a variance stabilising transformation to handle heteroscedasticity (Section 9.1.4). Here, we briefly introduce two contemporary concepts:

(1) *Variance modelling* using generalised least squares (GLS) models that allow the incorporation of variance functions to account for heteroscedasticity (function `gls` in R package *nlme*, Pinheiro et al. 2023).
(2) *Robust, heteroscedasticity-consistent covariance matrix estimation* (function `model_parameters` in R package *parameters*, Lüdecke et al. 2020).

We will use the 'anthocyanin' dataset to demonstrate these two approaches. Anthocyanin is a pigment that confers light protection in plants against UV radiation. This dataset contains measurements of anthocyanin concentrations in leaves of two woody plants species as a function of UV intensity (Figure 12.4).

```
## Read in data
   anthocyanin <- read.csv("anthocyanin.csv")

## Plot the data
> ggplot(data = anthocyanin, aes(x = uv, y = acn, colour = spec)) +
  geom_point() +
  labs(x = expression(paste("UV radiation (mW cm"^-2, ")")),
       y = expression(paste("Anthocyanin (", mu, "g cm"^-2, ")"))) +
    theme_test()
```

We first run a linear model (an analysis of covariance, see Chapter 11) including an interaction between UV radiation and plant species identity using the `lm` function.

```
## Ordinary least-squares regression model
> ols1 <- lm(acn ~ uv * spec, data = anthocyanin) # 'ols' stands for
ordinary least-squares

> summary(ols1)

Call:
lm(formula = acn ~ uv * spec, data = anthocyanin)
```

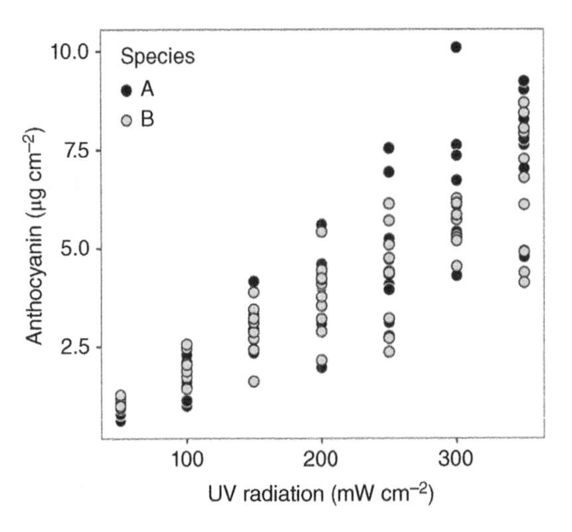

Figure 12.4 Foliar anthocyanin concentration of two plant species as a function of UV radiation. Note the increasing spread with increasing UV intensity indicating variance heterogeneity.

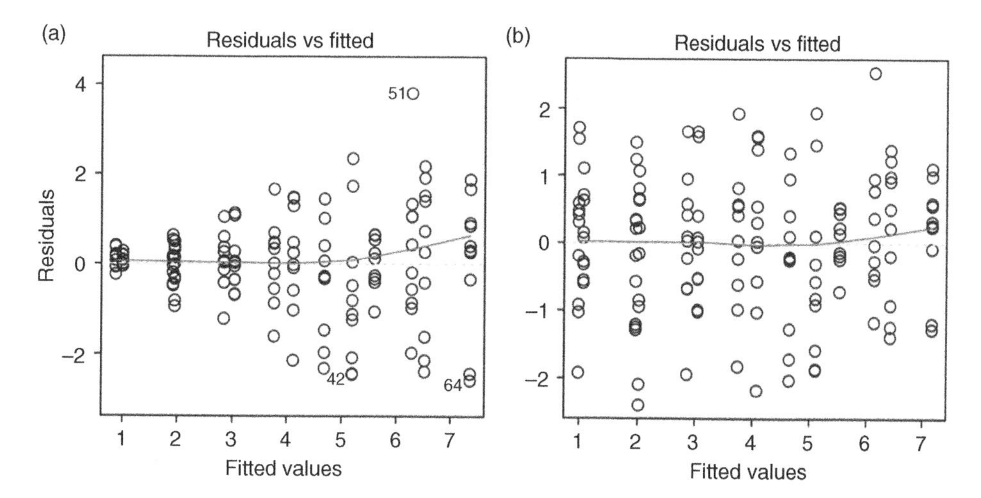

Figure 12.5 (a) Residual vs. fitted values plot associated with the ordinary least-squares model fitted with the `lm` function showing a fan-shaped pattern. (b) Residual vs. fitted values plot related to the generalised least-squares model fitted with the `gls` function, where a power variance function has successfully accounted for the heteroscedasticity as indicated by the equal spread of the residuals across the range of fitted values.

```
Residuals:
    Min      1Q  Median      3Q     Max
-2.5760 -0.4135  0.0683  0.4671  3.8143

Coefficients:
             Estimate Std. Error t value Pr(>|t|)
(Intercept) -0.206000   0.280559  -0.734    0.464
uv           0.021606   0.001255  17.220   <2e-16 ***
specB        0.291579   0.396770   0.735    0.464
uv:specB    -0.003229   0.001774  -1.820    0.071 .
---
Signif. codes:  0 '***' 0.001 '**' 0.01 '*' 0.05 '.' 0.1 ' ' 1

Residual standard error: 1.05 on 136 degrees of freedom
Multiple R-squared:  0.7911,    Adjusted R-squared:  0.7865
F-statistic: 171.7 on 3 and 136 DF,  p-value: < 2.2e-16
```

The model output indicates a borderline significant interaction between UV radiation and species identity.

For brevity, we focus on the residuals vs. fitted values plot in terms of model diagnostics. This residual plot shows a massive fan-shaped pattern, suggesting strong heteroscedasticity, so we cannot trust the model results (Figure 12.5a).

```
## Residuals vs fitted values plot
> plot(ols1, which = 1)
```

12.4.1 Variance Modelling Using Generalised Least-squares Estimation

The GLS approach is a generalisation of the ordinary least-squares technique. In generalised least-squares regression, the assumption of homoscedasticity (constant variance of errors) is relaxed, and the variance pattern of the errors is explicitly modelled. This is done by

Table 12.1 Available variance functions for variance modelling in generalised least-squares regression (R package *nlme*, Pinheiro et al. 2023).

Variance function	Description	Variance covariate (fitted values or predictor)
varFixed	Fixed variance	Continuous
varIdent	Different constant variances per stratum	Categorical
varExp	Exponential of the variance covariate	Continuous
varPower	Power of the variance covariate	Continuous
varConstPower	Constant plus power the variance covariate	Continuous
varConstProp	Two-component error model (additive and proportional error)	Continuous
varComb	Allows a combination of variance functions	Continuous and/or categorical

sub-models, so-called *variance functions*, that run under the hood and capture the error variance to then assign weights to the observations. The reciprocals of the estimated variances are used as weights, so that observations with smaller estimated variances weigh more heavily and can therefore contribute more information to the estimation process. At the same time, this translates into downweighting observations with higher variability. This makes the model more robust to unequal variance and ensures that statistical inference, such as the parameter estimates and especially their standard errors/confidence intervals and hypothesis tests, are not overly influenced by the presence of heteroscedasticity in your data.

Generalised least-squares models allow the errors to have unequal variance.

We will use the `gls` function (R package *nlme*, Pinheiro et al. 2023) to fit a generalised least-squares model. The variance functions are specified via the `weights` argument. There are six stand-alone variance functions available and an additional one that allows us to combine and apply multiple variance structures at the same time (Table 12.1).

We can run a `gls` model with and without variance functions on board. Without variance function, a `gls` model yields (nearly) identical results to a model fitted with `lm`. Slight differences that may occur are due to the fact that `gls` uses restricted maximum likelihood for parameter estimation, whereas `lm` relies on ordinary least-squares estimation.

Power and exponential variance functions are particularly well-suited for modelling fan-shaped residual patterns. Therefore, we now run a GLS model with a power variance function to deal with the heteroscedasticity issue.

```
## Generalised least-squares regression deals heteroscedastic errors by
variance modelling
> library(nlme)
> gls1 <- gls(acn ~ uv * spec, data = anthocyanin, weights = varPower())
```

The empty brackets of the variance function in the R code above default to `varPower(form = ~ fitted(.))` indicating that the fitted values of the model are used as variance covariate (just like in the residuals vs. fitted values plot). We can replace the fitted values with a continuous predictor (in our example data only UV intensity is available) and check if this yields even better results. We can use tools like the Akaike Information Criterion (AIC) (see Chapter 9) to compare models that have different variance functions but are otherwise identical.

Let us have a look at the `gls` model output, which is a little more comprehensive than the familiar `lm` output.

```
> summary(gls1)

Generalized least squares fit by REML
  Model: acn ~ uv * spec
  Data: anthocyanin
       AIC      BIC   logLik
  359.7459 377.2219 -173.873

Variance function:
 Structure: Power of variance covariate
 Formula: ~fitted(.)
 Parameter estimates:
  power
1.18242

Coefficients:
                  Value  Std.Error   t-value p-value
(Intercept) -0.04545116 0.08474437 -0.536333  0.5926
uv           0.02059613 0.00093744 21.970505  0.0000
specB        0.22118940 0.12175457  1.816682  0.0715
uv:specB    -0.00275446 0.00127869 -2.154131  0.0330

 Correlation:
         (Intr) uv     specB
uv       -0.815
specB    -0.696  0.567
uv:specB  0.597 -0.733 -0.809

Standardized residuals:
        Min          Q1         Med          Q3         Max
-2.40241073 -0.59988698  0.07910185  0.59101103  2.53712292

Residual standard error: 0.1826371
Degrees of freedom: 140 total; 136 residual
```

The GLS model results indicate that the UV × species interaction is statistically significant, but have we really managed to resolve the heteroscedasticity issue?

In the residuals vs. fitted values plot associated with the GLS model, the residuals scatter randomly around zero across the range of the fitted values indicating variance homogeneity (Figure 12.5b). This means that we can trust the model results and conclude that there is indeed a statistically significant interaction between UV radiation and species identity.

```
## Residual pattern after variance modelling
# plot(gls1) # returns a slightly differently formatted graph than in
Figure 12.5b
```

12.4.2 Robust, Heteroscedasticity-Consistent Covariance Matrix Estimation

The goal of this approach is to estimate a covariance matrix that is robust to the presence of heteroscedasticity. In other words, this technique aims to provide reliable standard errors and confidence intervals for the regression coefficients, even when there is unequal variance in the errors. This method is implemented using the `model_parameters` function (R package *parameters*, Lüdecke et al. 2020). We can choose a range of heteroscedasticity consistent

Hetero-scedasticity-consistent covariance matrix estimation gives robust standard errors in the presence of variance heterogeneity.

(HC) estimators running under the hood of the `model_parameters` function (see help file for `vcovHC`, R package *sandwich*, Zeileis et al., 2020; Zeileis, 2004). In the following example, we will apply the widely used HC3 estimator.

```
## Robust heteroscedasticity-consistent covariance matrix estimation
## Install and load required package
> install.packages("parameters")
> library(parameters)

## Extract the original coefficients, standard errors etc. from the lin-
ear model

> model_parameters(ols1) # same values as in summary(ols1) and in con-
fint(ols1)

Parameter       | Coefficient |       SE |         95% CI | t(136) |      p
---------------------------------------------------------------------------
(Intercept)     |       -0.21 |     0.28 | [-0.76,  0.35] |  -0.73 | 0.464
uv              |        0.02 | 1.25e-03 | [ 0.02,  0.02] |  17.22 | < .001
spec [B]        |        0.29 |     0.40 | [-0.49,  1.08] |   0.73 | 0.464
uv x spec [B]   |   -3.23e-03 | 1.77e-03 | [-0.01,  0.00] |  -1.82 | 0.071

> ## Compute heteroscedasticity-consistent standard errors

> model_parameters(ols1, vcov = "HC3")

Parameter       | Coefficient |       SE |         95% CI | t(136) |      p
---------------------------------------------------------------------------
(Intercept)     |       -0.21 |     0.19 | [-0.58,  0.17] |  -1.08 | 0.283
uv              |        0.02 | 1.36e-03 | [ 0.02,  0.02] |  15.93 | < .001
spec [B]        |        0.29 |     0.26 | [-0.23,  0.81] |   1.11 | 0.268
uv x spec [B]   |   -3.23e-03 | 1.84e-03 | [-0.01,  0.00] |  -1.76 | 0.081
```

The `t(136)` indicates a *t*-value with 136 degrees of freedom.

Comparing the two outputs (unadjusted and heteroscedasticity-adjusted), we can see that the model coefficients remained unaffected by the heteroscedasticity but the standard errors and thus confidence intervals of the intercept and the species effect were inflated by the heteroscedasticity as suggested by the considerably smaller values adjusted for heteroscedasticity. In our particular case, however, the conclusion about the significance of the interaction term remains unchanged with this method.

Box 12.2 Key to Linear Models and Extensions

Please note that we have included references to prospective models for frequently encountered data such as counts or rating scales, although these are not covered in this text. Modelling methods that do not appear in this book are marked with an asterisk (*).

1a. Response variable *continuous* **2**
(The response variable can theoretically take on any value within a specified range, e.g. length, mass, temperature, concentration)

NOTE: *Proportion or percentage* data is bounded by 0 and 1 (0 and 100%) and often derived from discrete variables and hence not considered truly continuous! (see *discrete* variables)

1b. Response variable *discrete* or strictly bounded (proportion/percentage) **7**
(Discrete variables can only take on certain values, e.g. counts, presence/absence, rating scales)

2a. Histogram of the continuous response variable **not** strongly right-skewed (i.e. no long tail to the right) **3**

2b. Histogram of the continuous response variable strongly right-skewed (i.e. shows long tail to the right) ▶ **Gamma Regression** `glm(..., family = Gamma)`*
or ▶ **Inverse Gaussian Regression** `glm(..., family = inverse.gaussian)`*

3a. Single explanatory variable **4**

3b. Multiple explanatory variables **5**

4a. Explanatory variable *categorical* (synonyms: factor, nominal variable, e.g. species identity, colour, type of rock, gender, blood type) ▶ **One-way Analysis of Variance (ANOVA)** `aov` or `lm`
→ See chapter 10

4b. Explanatory variable *continuous* ▶ **Simple Linear Regression** `lm`
→ See chapter 09

5a. All explanatory variables of the same kind **6**

5b. Mix of *categorical* and *continuous* explanatory variables (at least one of each kind) ▶ **Analysis of covariance (ANCOVA)** `lm`
→ See chapter 11

6a. Explanatory variables *categorical* ▶ ***n*-way Analysis of Variance (ANOVA)** `aov` or `lm`
→ See chapter 10

6b. Explanatory variables *continuous* ▶ **Multiple Linear Regression** `lm`
→ See chapter 09

7a. Response variable *discrete* **8**

7b. Response variable is a *proportion/percentage* (e.g. mortality, germination, proportion infested, percentage cover) **13**

8a. Response variable is **not** a *count* **9**

8b. Response variable is a *count* (whole numbers, integers) **10**

9a. Response variable *binomial/binary* (e.g. dead/alive, presence/absence, gender) ▶ **Logistic Regression** `glm(..., family = binomial)`
→ See chapter 12
(in simple cases **Chi-squared Test** → see chapter 7)

9b. Response variable *multinomial/ordinal* (e.g. rating scales).... ▶ **Ordinal Regression***
e.g. `clm` R package *ordinal*

10a. Count data not zero-inflated **11**

10b. Zero-inflated count data **12**

11a. Overdispersion absent ▶ **Poisson Regression** `glm(..., family = poisson)`*

11b. Overdispersion present ▶ **Negative Binomial Regression** `glm.nb` R package *MASS**

12a. Overdispersion absent ▶ **Zero-inflated Poisson Regression** `zeroinfl` R package *pscl**

(Continued)

Box 12.2 (Continued)

12b. Overdispersion present ▶ **Zero-inflated Negative Binomial Regression***
`zeroinfl(..., dist = "negbin")` R package *pscl*

13a. Proportions/percentages based on counts (underlying count data is available, e.g. 10% mortality = 10 dead individuals out of 100). ▶ **Logistic Regression** `glm(..., family = binomial)`

13b. Proportions/percentages lacking the underlying counts (e.g. estimates of percentage vegetation cover, cloud cover, body fat) ▶ **Beta Regression** `betareg` R package *betareg**
or ▶ **Linear Regression** after logit transformation of the response variable `lm`*

References

Baty, F. and Delignette-Muller, M.L. (2014). nlsMicrobio: Data Sets and Nonlinear Regression Models Dedicated to Predictive Microbiology. *R Package Version 0.0-1*. https://CRAN.R-project.org/package=nlsMicrobio.

Baty, F., Ritz, C., Charles, S. et al. (2015). A toolbox for nonlinear regression in R: the package nlstools. *Journal of Statistical Software* 66 (5): 1–21. https://doi.org/10.18637/jss.v066.i05.

Canty, A. and Ripley, B. (2022). boot: Bootstrap R (S-Plus) Functions. *R Package Version* 1.3-28.1. https://CRAN.R-project.org/package=boot.

Davison, A.C. and Hinkley, D.V. (1997). *Bootstrap Methods and Their Applications*. Cambridge: Cambridge University Press, 582 pp. ISBN 0-521-57391-2, https://doi.org/10.1017/CBO9780511802843.

Jackman, S. (2020). *pscl: Classes and Methods for R Developed in the Political Science Computational Laboratory*. Sydney, New South Wales, Australia: United States Studies Centre, University of Sydney R package version 1.5.5.1. https://github.com/atahk/pscl/.

Lüdecke, D., Ben-Shachar, M., Patil, I., and Makowski, D. (2020). Extracting, computing and exploring the parameters of statistical models using R. *Journal of Open Source Software* 5 (53): 2445. https://doi.org/10.21105/joss.02445.

Miguez, F. (2023). nlraa: nonlinear regression for agricultural applications. *R Package Version* 1 (9): 3. https://CRAN.R-project.org/package=nlraa.

Onofri, A. (2020). The broken bridge between biologists and statisticians: a blog and R package. *Statforbiology, IT*. https://www.statforbiology.com.

Pinheiro, J., Bates, D., and R Core Team (2023). nlme: linear and nonlinear mixed effects models. *R Package Version* 3: 1–162. https://CRAN.R-project.org/package=nlme.

Ritz, C. and Streibig, J.C. (2008). *Nonlinear Regression with R*. New York, NY: Springer New York.

Ritz, C., Baty, F., Streibig, J.C., and Gerhard, D. (2015). Dose-response analysis using R. *PLoS One* 10 (12): e0146021.

Sit, V. and Poulin-Costello, M. (1994). *Catalogue of Curves for Curve Fitting. Biometrics Information Handbook Series No. 4*. British Columbia, Canada: Forest Science Research Branch, Ministry of Forests.

Wickham, H., Averick, M., Bryan, J. et al. (2019). Welcome to the *tidyverse*. *Journal of Open Source Software* 4 (43): 1686. https://doi.org/10.21105/joss.01686.

Wickham, H. (2023) Differences between the base R and magrittr pipes. https://www.tidyverse.org/blog/2023/04/base-vs-magrittr-pipe/.

Wood, S.N. (2011). Fast stable restricted maximum likelihood and marginal likelihood estimation of semiparametric generalized linear models. *Journal of the Royal Statistical Society (B)* 73 (1): 3–36.

Wood, S.N. (2017). *Generalized Additive Models: An Introduction with R*, 2nd ed. Chapman and Hall/CRC, 496 pp.

Zeileis, A. (2004). Econometric computing with HC and HAC covariance matrix estimators. *Journal of Statistical Software* 11 (10): 1–17. https://doi.org/10.18637/jss.v011.i10.

Zeileis, A., Köll, S., and Graham, N. (2020). Various versatile variances: an object-oriented implementation of clustered covariances in R. *Journal of Statistical Software* 95 (1): 1–36. https://doi.org/10.18637/jss.v095.i01.

Index

References to Figures are in *italics*, to tables in **bold**, and R functions in `courier` font.

R-ticulate: A Beginner's Guide to Data Analysis for Natural Scientists, First Edition.
Martin Bader and Sebastian Leuzinger.

Companion website: www.wiley.com/go/Bader

u

v

w

Printed and bound by CPI Group (UK) Ltd, Croydon, CR0 4YY

16/04/2025

14658370-0003